Aspects of Time

Aspects

Hackett Publishing Company

of Time

GEORGE N. SCHLESINGER

Indianapolis • Cambridge

Library of Congress Catalog Card Number 79-66954

ISBN 0-915144-69-7
0-915144-70-0 (pbk)

Cover design by Laszlo J. Balogh
Interior design by James N. Rogers

For further information, please address
Hackett Publishing Company, Inc.
P.O. Box 55573
Indianapolis, Indiana   46205

Contents

CONTENTS

Acknowledgments

I wish to thank the Editors of the *Australian Journal of Philosophy, British Journal for the Philosophy of Science, Mind, Philosophical Quarterly, Philosophical Studies,* and *Review of Metaphysics* for having permitted me to use their journals in which to advance some of the ideas which are explored more thoroughly in the present book.

Introduction

Although the philosophy of time is perhaps the most fascinating area in the whole of philosophy, it may present to those who encounter it for the first time some unexpectedly thorny problems. But once these initial trouble spots are mastered, students usually develop a deep and lasting interest in the subject. On a superficial look its difficulties appear surprising. Why should a question such as whether or not time flows, and events of the future approach us and momentarily become present events after which they recede into the past, be so perplexing? After all, doesn't it deal with a feature of the universe which we experience every moment of our lives? Apparently, it is the very proximity of the phenomenon that interferes with our focusing clearly on it. What proves to be so absorbing about the philosophy of time is that even when we penetrate its familiar components, the character of its core remains enigmatic.

This work is devoted exclusively to the examination of the metaphysical aspects of time. These are empirical features with which physicists do not concern themselves, preferring to leave their investigation to philosophers. The student may be relieved to learn that no knowledge of physics or mathematics is required for the complete understanding of any part of this book.

The fact that the philosophy of time is generally found by the uninitiated to be more elusive than other topics in philosophy— such as the philosophy of mind or epistemology—may account for its comparative neglect. My small contribution to the subject consists in attempting to make what I have to say as widely accessible as possible. Thus, I have opted to forego presenting in the body of the text every argument with all the necessary qualifications, nor the various possible objections and queries and the ways of meeting these; instead, I have placed these materials in a separate section.

By going through the first part of the book, the reader will gain a general impression of the structure of the major arguments. At this point he will probably be left with a number of unanswered questions, some of which will be raised, along with others, in the second section. There are nearly sixty such questions and objections, to which I provide what appear to me as the most adequate answers. Ideally, the reader should try to answer these questions and objections before reading the answers provided. Permit me to emphasize the importance of the questions and answers in illuminating the topics dealt with in the text. For this reason, I hope no reader will be tempted to skip them.

I do not claim, by any means, to have achieved maximum clarity in this work. But by not forcing the reader to ingest all the material in a single dose and by presenting part in the form of questions and answers, I hope to make the reader's task considerably less arduous than it otherwise might be. I have marked within the text the places at which it may be most useful to consult the Question and Answer sections—e.g., (v. Q&A 15)—with a view of obtaining often indispensable additional discussion of the issues.

I

THE SIMILARITIES
BETWEEN TIME AND SPACE

1 Some Obvious Similarities

Time and space are the most fundamental features of the physical universe. Everything that is exists *in* time and *in* space. Time and space are containers, everything is located in them; every event occurs *at* some point in time and *at* some point in space. Particulars exist which lack all physical properties except the most basic ones: temporal and spatial properties. For instance, whereas material objects must be of this or that temperature, of this or that mass, of this or that electrical potential, an event such as 'the light is changing from green to red' cannot be said to possess any of these properties to any degree. Yet the same event still possesses the temporal property of happening at a given time and the spatial property of happening at a given place.

Because of the unparalleled ubiquity of space and time which sets them far apart from anything else, it is not surprising that people tend to believe that the two are fundamentally similar. Various philosophers have expressed the opinion that space and time share common basic properties. I believe it is a good idea to choose as our first topic the question of the similarity of space and time, for discussing it will throw light on some very fundamental issues. For a start we may consider what Nelson Goodman has said on this subject:

> . . . the analogy between space and time is indeed close. Duration is comparable to extent. A thing may vary in color in its different spatial or its different temporal parts. A thing may occupy different places at one time or the same place at different times, or may vary concomitantly in place and time. The relation between the period of time occupied by a thing during its entire existence and the rest of time is as fixed as the relation between the region the thing covers during its entire existence and the rest of space. And yet there is this difference: two things may approach and then recede from each other in space, they grow more and then less alike

3

in color, shape, etc; but two things never become nearer and then further apart in time. The location or the color or the shape of a thing may change but not its time.[1]

This passage is especially useful because it contains an elementary error which illustrates how easy it is to fall into confusion when comparing time and space. When questioning whether a given property of space has its exact parallel among the properties of time, it is essential that we make sure we get hold of a feature of time which is truly the temporal counterpart of the feature of space under examination.

In order to see quickly Goodman's error, let us look at the world lines of two particles, α and β. I have promised to keep our discussion entirely nontechnical, and I believe that by introducing the device of a 'world line', whereby the history of an individual may be illustrated graphically, rather than making things more difficult through technicalities, I render them in a more comprehensible manner by making them visualizable.

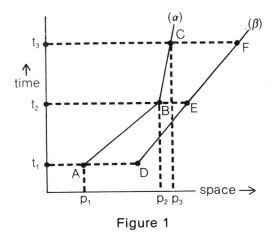

Figure 1

Particle α is at p_1 at t_1, from where it travels at uniform velocity to p_2, arriving there at t_2. The fact that AB is a straight line shows that the velocity between p_1 and p_2 is uniform. At p_2 the particle abruptly slows down and travels at a lower uniform velocity to p_3. The fact that BC is less inclined to the space axis than AB shows that less space is covered in more time, or that the velocity between p_2

1. *Problems of Space and Time* ed. J. J. C. Smart. New York, Macmillan, 1964, pp. 367–8.

and p_3 is lower than between p_1 and p_2. On the other hand, β travels at a uniform velocity between time t_1 and t_3. Focusing attention on the horizontal lines AD, BE and CF, it becomes evident that α and β approach first and then recede from one another in space, for it is

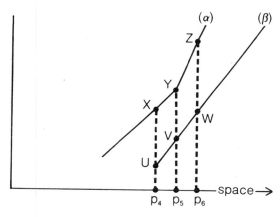

Figure 2

seen that BE, which is the spatial distance between α and β at t_2 is smaller than AD, which is the spatial distance between them at t_1; whereas CF, which is the spatial distance between the two at a yet later time, t_3, is again greater than BE.

But then exactly in parallel fashion from Figure 2, it is obvious that α and β do in fact approach one another at first but then recede from one another in time. At p_4 the temporal distance between the two particles is clearly given by the vertical distance XU, which is greater than the vertical distance YV representing the temporal distance between α and β at p_5. However, at a still further place, p_6, the temporal distance between them increases once more, as shown by the fact that the vertical distance ZW is greater than YV.

Of course the notion of 'the spatial distance between α and β at t' is a familiar one, which plays a crucial role in many practical problems. For example, the magnitude of the force of gravitational attraction between α and β at t depends on the spatial distance between α and β at t. The notion of 'the temporal distance between α and β at p' has no such significance; there is no known functional relationship between that distance and some important physical property. This may well be the reason it escaped Goodman's attention. It is obvious, however, that this is the relevant notion to our inquiry whether things may become nearer and further apart in time as well as in space. The magnitudes XV, YV and ZW and their relationships are the exact temporal counterparts of the spatial distances AD, BE and CF and the relationships among them.

It may be countered however that although physical objects definitely can become nearer as well as further apart in time, our analysis nevertheless shows that space and time are not entirely similar. After all, it had to be admitted that 'the spatial distance between α and β at t' is a significant notion which plays a substantial role in determining some significant physical relations between α and β, whereas 'the temporal distance between α and β at p' has nothing significant associated with it to be worthy of our attention.

In answering this objection, we must concede that space and time are not exactly alike in every respect. The Doctrine of the Similarity of Space and Time, if it is to be maintained at all, can only be done so with respect to the necessary features of space and time. The Doctrine would then amount to saying: if we have a statement about space that is necessarily true or necessarily false, then the temporal counterpart of that statement must also be necessarily true or false, respectively. Now it may be maintained correctly that it merely happens to be true that the spatial distance between α and β determines a number of physical relations between α and β, but there is no conceptual necessity that it be true. Hence, the Doctrine does not apply to it, and the temporal distance between α and β may not be relevant to any of the physical relations between α and β. On the other hand, the statement, 'Things may get nearer and then further apart in space,' seems to be true by conceptual necessity. Therefore the Doctrine applies to it. (v. Q&A 1, 2, and 3)

2   *The Question of the Scope of the Doctrine of the Similarity of Space and Time*

Richard Taylor has made a major contribution toward a detailed defense of the Doctrine of the Similarity of Space and Time. In a fine article called "Spatial and Temporal Analogies and the Concept of Identity" he claims that

> . . . many propositions involving temporal concepts which seem obviously and necessarily true are just as necessarily but not obviously true when formulated in terms of spatial relations.[2]

He then goes on to examine a number of temporal propositions which are necessarily true, yet whose spatial counterparts do not

2. Op. Cit. p. 381.

seem to be true, thus apparently representing cases in which space and time do not resemble one another. Taylor is successful however in showing in each case that the lack of parallelism is illusory. He ably demonstrates in each instance that the proposition we thought to be the spatial counterpart of the original true temporal proposition is in reality no such counterpart. He then carefully constructs what is indeed the genuine spatial counterpart of the given true temporal proposition on, and this turns out to be in fact true.

There are however two very important questions that Taylor leaves unanswered. The first question is which true temporal propositions do have true spatial analogues and which do not; or, in what kind of properties are space and time similar to one another and in what kind of properties are they not? Taylor claims only that 'many propositions involving temporal concepts which are true do in spite of appearances have corresponding true propositions involving spatial concepts,' but not all. But it is of great interest for us to know what characterizes those true temporal propositions to which true spatial propositions do correspond, and those to which such propositions do not correspond. We should be able to answer the question, 'by what criterion are propositions involving temporal concepts divided into the class the members of which remain true when formulated in terms of spatial relations, and the class the member of which cease to be true when so formulated?' If a demarcation principle between the two types of propositions exists, and we are able to come up with a characterization of those temporal statements that do have symmetrical counterparts, then it is indeed of interest to have a close look at any true temporal statement that has the characteristics of a statement which ought to have a true spatial counterpart but does not seem to have one. Enough motivation then would be provided for making efforts to show that in a correct analysis such a counterpart does exist. But if there is no systematic way in which temporal propositions may be divided into two kinds, and it just so happens that some temporal propositions remain true when reformulated in spatial terms and others do not, then I can see little point in making elaborate efforts to show that some true temporal propositions, in spite of appearances to the contrary, have true spatial counterparts.

The second question is why all those true temporal propositions have true spatial counterparts as well. If all such propositions can be recognized by their possession of certain characteristics, it is of great interest to know why such possession is sufficient to assure the existence of true spatial counterparts. Whatever the reason, it is to be expected that our knowledge of it should increase our understanding of the nature of space and time. The need to answer the first question would of course be entirely obviated if we adopted the position that the Doctrine of the Similarity of Space and Time applies without any qualification to all necessary statements about space and time. In view of the fact that Taylor has shown that all seeming counterexamples cited by him vanish on correct analysis, in addition to which James W. Garson[3] has shown that some other candidates do not constitute counterexamples either, and furthermore that no one so far has succeeded in establishing the existence of a counterexample, such a position may seem reasonable. Yet if Doctrine was indeed unqualifiedly true, it would be most puzzling. For the concept of time is an entirely distinct concept from the concept of space. Spatial properties present themselves in entirely different ways to our senses; they are perceived differently. It is true, for instance, that we have both temporal distances and spatial distances, but they differ vastly from one another, with one being measured by instruments and methods that are entirely dissimilar from those by which the other is measured. The same applies to all other temporal properties having spatial counterparts. There appears to be no good reason why temporal properties should obey the same or even similar laws as spatial properties.

The purpose of this chapter is to show a number of things. First of all I shall show that the Doctrine of Similarity of Space and Time does not apply to all propositions; that there are necessarily true temporal propositions which do not have true spatial counterparts, and vice versa. I shall also show that there is an easy and obvious method to characterize those true temporal propositions which do have symmetrical counterparts. It will also become evident that there is a compelling reason why all those true temporal propositions which have symmetrical counterparts must have them.

3   *Apparent Counterexamples to the Doctrine*

Before being able to establish to what extent space and time are similar, we must know how to determine whether a given true temporal statement has a true spatial counterpart. In order to achieve this we must make sure that we can always identify the right statement into which a temporal statement must be turned to yield its spatial counterpart. The general principle to follow is that when a statement asserting something about time makes also a reference to space, to identify the relevant counterpart, we have to find a statement making a parallel assertion about space and parallel reference to time. In other words, to produce the spatial counterpart of a

3. James M. Garson, "Here and Now," *The Monist* (1969), pp. 469–477.

temporal statement, we must change every temporal term into the corresponding spatial term and every spatial term into the corresponding temporal term.

In this section I shall consider a number of examples which seem to show that space and time are dissimilar. However, on careful scrutiny we shall find that they turn out not to constitute real counterexamples to the Doctrine of the Similarity of Space and Time. These examples illustrate the various pitfalls to be avoided when trying to assess the validity of the Doctrine.

(1)   The following example illustrates that it may not be always so obvious how to change every temporal term into the corresponding spatial term and vice versa, because we may often overlook terms which do not refer overtly to temporal or spatial predicates yet do have such predicates hidden in them. An error of translation can be committed easily through a failure to recognize a term as being temporal or spatial, and, therefore, as being in need of inversion in order to produce a complete translation of one kind of statement into its exact counterpart:

> A further disanalogy between 'here' and 'now' which shows that there is no spatial analogy to temporal becoming is due to there being no spatial analog to *waiting*. If I wait for n-time units my use of 'now' denotes a different time regardless whether I move or not. But the spatial analog of this is nonsense since there is no sense to 'waiting for or through space'.[4]

The temporal statement under review in this case is 'If I wait n-time units my use of 'now' denotes a different time regardless whether I move or not', and the presumed spatial counterpart is 'If I wait n-space units my use of 'here' denotes a different place regardless whether I move or not'. The latter statement not only fails to be true but is even devoid of meaning. The failure of the true temporal statement under review to translate into a true spatial statement shows in Gale's opinion that space and time are radically different.

But before the question of the existence of a true analogue can be settled, we must make sure that we are doing the translation correctly, and this cannot be the case unless we realize that the terms 'wait' and 'move' have temporal and spatial terms hidden in them which have to be made explicit and be inverted properly. The term 'wait n-time units' in the first statement stands for 'occupy temporal positions t_1 through t_n', and 'move' stands for 'shift my

4. R. Gale, "'Here' and 'Now'", *The Monist* (July, 1969), p. 409.

spatial position'. Accordingly, the first statement, with all its temporal and spatial terms laid bare, runs 'If I occupy temporal positions t_1 through t_n, then my use of 'now' at t_1 and at t_n denote different times, regardless whether I shift my spatial position; that is, regardless whether I make the two utterances at the same place or not'. The exact spatial counterpart of this statement is 'If I occupy spatial positions p_1 through p_n, then my use of 'here' at p_1 and p_n denote different places regardless whether I shift my temporal position; that is, regardless whether I make the two utterances at the same time or not'. This second statement is, of course, as true as the first one. There is therefore no disanalogy between space and time with respect to the relations referred to in the two statements.

(2) Following is another example in which it may seem that space and time are disanalogous and in which it is not at once clear what mistake of translation had been committed:

'If the birth and death of A and B coincide in time, then every time occupied by A is also occupied by B, and vice versa'. This statement is true. The spatial counterpart seems to be:

'If the birth and death of A and B coincide in space, then every place occupied by A is also occupied by B and vice versa'. This statement, however, is obviously false.

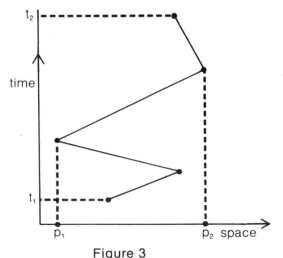

Figure 3

The mistake can be uncovered nonetheless once we realise that the spatial counterpart of 'to be born at t_1' is not 'to be born at p_1'. The phrase 'to be born at t_1' means 'to have t_1 as the most extreme point in time occupied by an individual' (for if an individual is born at t_1, then it follows that he does not extend temporally in one direction beyond t_1) but 'to be born at p_1' does not mean 'to have p_1 as the most extreme point in space occupied by an individual'. Bearing this in mind we realize that the correct counterpart of our temporal statement would run along the following lines: 'If p_1 is the most extreme point in one direction, beyond which there is no point which is occupied by either A or B at any time, and p_2 is the most such extreme point in the other direction, then (in one-dimensional space) every place occupied by A at one time or another is also occupied by B at some time, and vice versa.'

It so happens that there is no word in English denoting the spatial equivalent of being born at t_1 and dying at t_2. We could of course invent one, such as 'sborn at p_1' and 'sdying at p_2'. Figure 3 illustrates the world line of a given particle of which it is true to say that it is born at t_1, dies at t_2, is sborn at p_1, and sdies at p_2. Obviously, it is true also that if the sbirth and sdeath of A and B coincide in space, then every place occupied by A is also occupied by B, and vice versa.

(3) It must be noted that everyone will acknowledge that there is a certain disanalogy between space and time, but it is of a kind which cannot serve as a counterexample to the thesis that space and time are radically alike. The temporal statement 'If we know that an object occupies two nonadjacent points in time, t_1 and t_2, and is continuously extended temporarily, then we can name other points in time which it also occupies' is true. However, its spatial counterpart 'If we know that an object occupies two nonadjacent points in space, p_1 and p_2, and is continually extended spatially, then we can name other points in space which it also occupies' is false. This lack of analogy is brought about by the fact that time is one-dimensional and space is multi-dimensional. Consequently, if an object occupies t_1 and t_2, it occupies all the temporal positions which exist between t_1 and t_2. But if an object occupies p_1 and p_2, it does not necessarily occupy all the points between them—only all the points on one of the infinitely many lines which connect p_1 and p_2. In order to show that space and time are radically different, we would have to show that a necessarily true temporal statement cannot be transformed into a true spatial statement, even in one-dimensional space. (v. Q&A 6)

(4) When we look at extended particulars that have both spatial and temporal parts, it appears that they are fundamentally different from one another. We may consider a complex piece of machinery such as a television set, which has both spatial and temporal extension. Suppose we chopped off several years from its ex-

istence. Doing so would not radically interfere with the nature of the remaining temporal parts of the set. All we would have done is produce a TV set with a shorter life span, but the remaining temporal part of the set would function as properly as the original, temporally untruncated set. Suppose, on the other hand, that we reduced the amount of space occupied by the set by removing its entire front part. This act would certainly interfere with the nature of the remaining spatial part. The part of the set which would now be left to us would not simply be a television set of smaller spatial dimension, but a useless wreck. It is obvious, therefore, that the different spatial parts of a TV set hang together and are mutually dependent on one another in a way in which its different temporal parts are not. The same is true of innumerable other physical systems such as cars, ships, clocks or, for that matter, supermarkets, hospitals, air terminals and so on, as well as most animals and plants. Breaking them into their spatial components yields a number of objects very different from the original whole, which was a complex functioning system. On the other hand, when these are divided into their component temporal parts, the resulting systems are merely shorter versions of the original whole.

I believe, however, that the correct position is not to interpret this phenomenon as a manifestation of a fundamental difference in the nature of space and time themselves. Rather, the difference should be looked on as a consequence of the peculiar nature of a complex mechanical system such as a television set: temporally, it consists of a series of more or less identical parts, each constituting a full-fledged set. Spatially, however, it does not constitute an assembly of identical parts; when combined together, its various parts have different properties, each of which contributes its unique share to accomplishing whatever the complex whole is supposed to do.

This answer may not seem acceptable at once, for it might well be asked: in cases where a given temporal statement is true about a set of particulars while the spatial counterparts of these statements is false, how do we know whether this difference is to be taken as a sign of a difference between the temporal and spatial properties of the given set of particulars or as a sign of a difference between time and space itself? Is there really more involved than an arbitrary decision? I believe, however, that this objection is easily met: if all particulars, without exception, were different with respect to a particular temporal property and its spatial counterpart, then we would have compelling reason to assign the difference to space and time themselves. But such is not the case when only some particulars exhibit this difference, while others do not. In the present case the asymmetry we found should not be assigned to space and time themselves because there happen to exist many physical particulars which are exactly conversely asymmetrical with respect to space and time such as a particular like a TV set. The existence of two kinds of particulars—particulars whose remaining parts are left unaffected when truncated temporally but whose remaining parts are radically different from the previous whole system when truncated spatially; and another kind of particulars for which the loss of a temporal part makes a crucial difference but not the loss of a spatial part—indicates decisively that we have not found here a fundamental disanalogy between space and time. Rather, we may conclude that there is a set of particulars that show a basic asymmetry with respect to space and time, but that this set has its counterpart whose members show just the opposite kind of asymmetry. A counterpart in the relevant sense to a particular such as a complex machine would be, for example, a symphony heard throughout an extended region of space and time. Suppose we reduced the space throughout which the symphony could be heard—this would leave us with the same symphony, but one heard in fewer places. It would be quite different, however, if we chopped off a considerable temporal part of the same symphony, say one-half of each movement. We would be left not merely with a shorter symphony but with something that was not musically coherent—something entirely different from the original symphony. Were it not a fact that we could produce such a physical counterpart to a mechanical system in which the asymmetry is reversed, and were it the case that all particulars had essentially the features of television sets with respect to temporal and spatial truncation, then we might have well been forced to agree that there is a basic difference in the nature of time and space themselves, but not under the prevailing circumstances.

(5) The following plausible-looking counterexample to the Doctrine of the Similarity of Space and Time has been produced by Ian Hinkfuss, a philosopher who seems to have given the matter considerable thought. In a section of his book titled 'More Differences Between Time and Space' he says

> . . . a glance down the list of properties of space would convince many people (if they needed convincing) that time is very different from space.[5]

Among such properties he mentions transparency to light, which is

---

5. *The Existence of Space and Time* (Oxford 1975), p. 80.

a quality that a given section of space may have but not a given section of time.

At first look it might seem that Hinkfuss has really hit on something substantial, that is, he has discovered not merely some contingent difference but a more far-reaching one. After all, he seems to be right in thinking that it is not merely that a given interval of time does not happen to be transparent to light but altogether it does not appear to make sense to speak of time as being or not being transparent to light. Thus, it is analytically ensured that it is not the case that a certain time interval is transparent to light. Similarly, it is analytically or necessarily true that it is not the case that a given time interval is opaque to light.

In spite of the initial plausibility of what Hinkfuss says, on further reflection it becomes evident that it is merely a contingent feature of the English language that one may use the words 'transparent' and 'opaque' to light in application to spatial intervals but not to temporal intervals. It is easily seen however that in principle there is no conceptual barrier in applying these terms to temporal intervals. This is best illustrated once more by space-time diagrams. In Figure 4 the various straight lines depict the worldlines of light-wave fronts. Because these lines all stop at p_1 and begin again at p_2, we say that the spatial interval $p_1 - p_2$ is opaque to light, whereas that of $p_2 - p_3$ is transparent. But then entirely

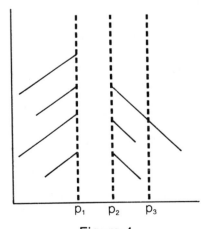

Figure 4

analogously, considering Figure 5 the temporal interval, $t_1 - t_2$ is opaque whereas the temporal interval $t_2 - t_3$ is transparent. We now realise that the dissimilarity is merely contingent. We have in fact encountered spatial intervals like $p_1 - p_2$ but not temporal intervals like $t_1 - t_2$. Because of our different experiences with spatial and temporal intervals, the terms 'transparent' and 'opaque' have an established usage with respect to spatial intervals only. (v. Q&A 5)

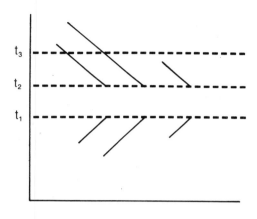

Figure 5

4   *Genuine Counterexamples to the Doctrine*

We have considered various seeming counterexamples to the Doctrine of the Similarity of Space and Time and found, on examining them carefully enough to avoid the subtle pitfalls into which investigators of our topic may fall, that these constitute no counterexamples. Numerous more such cases are presented in the existing literature. After such consideration, it is not unnatural to want to conclude, as have a number of philosophers, that the Doctrine may be unchallenged. But such a conclusion would be false. There are numerous counterexamples to the Doctrine. I shall cite a few and discuss some very briefly. But in order to establish conclusively the falsity of the Doctrine, a single clear-cut counterexample will suffice, as follows:

(1)   The following statement will universally be regarded as true:

(S)   The fact that two events, E_1 and E_2, occurred at the same *place* cannot be known for certain unless E_1 and E_2 also occurred at the same *time*.

We may of course arbitrarily designate a point on earth or on some celestial body as the origin of our spatial coordinate system, but apart from that there is no conceivable way to ascertain whether E_1 and E_2, which occurred at different times, really occurred at the same place. Some would go even further and say that not only can we not secure an answer to this question but that the question is altogether devoid of meaning. But even those who subscribe to the idea of an 'absolute space' and hence maintain that it makes sense to ask whether E_1 and E_2 did or did not occur at the same place agree that we have no way of answering this question. It is different with the temporal counterpart of (S):

> (T)   The fact that two events, E_1 and E_2, occurred at the same *time* cannot be known for certain unless E_1 and E_2 also occurred at the same *place*.

(T) is of course not necessarily true. Without arbitrarily designating any event as the origin of our temporal coordinates, it is in principle possible to ascertain the simultaneity of two distant events provided we may assume that signals marking their occurrence travel instantaneously to the observer. Before proceeding, the following question should be raised: does (S) express a necessary truth? And the answer is no. If, for example, light did not travel at constant speed relative to all objects but slower relative to those which moved in the same direction and faster relative to those which traveled in the opposite direction, then many would agree that it could be detected whether an object was at absolute rest, and therefore (S) would be false. But if the principle of relativity, according to which neither through sending out light signals or by any other means is it possible to determine the velocity of a given object relative to space itself, is true, then (S) is necessarily true. (T) on the other hand is true only if there is a law of nature that there are no infinitely fast signals. The conditions that render (S) true are not related in any manner to the condition which renders (T) true. To determine the co-temporality of events, infinitely fast signals are required; whereas to determine their co-spatiality requires the falsity of the principle of relativity. The first requirement is certainly not a translation of the second requirement into temporal terms; the two requirements are by means counterparts of one another. Admittedly, we are dealing here with differences in the ways in which given occurrences are ascertained to have taken place and therefore, in a certain sense, we are dealing here with epistemic differences. However, this cannot be put down as merely differences in the ways in which our perceptive organs are oriented toward space and time. The reason we achieve knowledge of co-temporality so differently from the way in which we arrive at knowing co-spatiality is rooted in the basic difference between the way in which the position of an event is fixed in the temporal series of moments and the way in which the position of an event is fixed in the spatial sequence of places. (v. Q&A 7, 8, and 10)

(2)   It seems for a variety of reasons that whereas it is true that temporal positions are ordered, and thus time has a direction, space has no direction. I shall indicate briefly but two possible ways in which one may argue that with respect to directionality, a fundamental difference between space and time exists.

(a)   Given that in a typical temporally developing system, such as an animal, in which its state of infancy is at t_1 and its state of maturity is at t_2, at once it can be deduced that for all the indefinitely many similar systems in the world—such as all other animals, all plants, and even inorganic systems such as buildings and machines—their younger and older stages (where 'young' and 'old' are not defined in terms of age but it terms of chemically relevant features associated with age) are also oriented in the direction from t_1 to t_2. Nothing similar exists with respect to space. We may, for example, know that the head of a given animal is at p_1 and its tail at p_2, but this does not enable us to deduce the direction in which any other animal is facing. It would require an extensive discussion to establish firmly that the posession of directionality is an intrinsic feature without which time would lose its very essence; that is, the possession of order is *necessarily* true of time.

(b)   Causal influence proceeds from earlier to later in time. In the fourth chapter I shall introduce arguments to support the contention that backward causation is logically impossible. According to these arguments when two events, E_1 and E_2, are causally connected, and E_2 is dependent on E_1 in a way in which E_1 is not dependent on E_2, and consequently we have to regard E_1 as being the cause of E_2, then it is logically impossible that E_2 should precede E_1. Thus the temporal relations of 'earlier' and 'later' are logically connected with the direction of causation. There is no parallel relation that is similarly connected with direction in space.

(3)   Everything that has a position in time is temporally related to everything else in time. This phenomenon has sometimes been referred to as the 'unity of time'. Space, however, does not necessarily have such unity. The spatial entities of one given space could be entirely unrelated to those belonging to another space. These would be then entirely disparate spaces. On the other hand, disparate times are inconceivable.

For a detailed argument in support of this thesis the reader is referred to A. Quinton's "Spaces and Times" in *Philosophy* (1962), pp. 130–147. Quinton considers the space–time system of a single individual. The same individual may divide his life and live alternately in two spaces which are not at all related. A point in the first space cannot be connected by any line to any point in the second space. But the same person cannot live alternately in two times which are similarly unrelated. A moment in one time system is always either before or after any moment in the second time system. Some have disputed Quinton's claim, but it seems to me that his thesis can be defended successfully. But a defender of the Doctrine of the Similarity of Space and Time is disadvantaged by having to commit himself to the specific assumptions made by the opponents of Quinton. (v. Q&A 9)

(4)   In his book *Individuals* P. F. Strawson elaborately constructs a world that has become a frequent subject of philosophical discussion, a world in which nothing exists except sounds of different timbre, pitch and volume. This is a one-dimensional world, in which the origin is marked by a master-sound and all other sounds are of a fixed distance from the origin; where 'distance' between the origin and given sound is not the amount of space between them, for this is a no-space world, but is defined uniquely in terms of a quantity measurable in this universe. Strawson shows in detail that such a world, despite the fact that space does not exist at all, is rich enough to allow us to travel forward and backward and reidentify the different individual sounds that are dispersed throughout this universe and which constitute the particulars, besides us, which inhabit it. There has been no attempt to prove this, but all readers of Strawson's account are likely to conclude that it would not be possible to construct a parallel world in which space existed but time did not. In other words, a temporal world devoid of space would not be too impoverished to sustain a system of particulars of a certain kind; whereas a spatial world in which time did not exist would be entirely stripped of its capacity of containing the basic ingredients of a viable universe. This shows that time is a more fundamental element of the world than space, which we could do without and still construct conceptually a fully describable, stable world. However, to remove time from our conception would be to destroy everything.

5   *What Time and Space Have in Common and Aspects in Which They May Differ*

The question as to which class of properties the Doctrine does apply and to which it does not is really not very difficult to answer.

Suppose it were asked whether there was a symmetry between the properties of a dog and a table. The answer obviously is that with respect to a certain class of properties, a dog and a table are necessarily entirely similar but not so with respect to any property outside that class. A dog as well as a table is a material object; thus the two resemble each other perfectly with respect to those properties that all material objects have in general. Newton's three laws of motion, for instance, apply to dogs no less than to tables, because those laws are obeyed by all material objects. Yet a dog has many properties which uniquely contribute to it being specifically a dog and not some other material object, and there is no reason why it should share these properties with a table. At first it may seem bizarre to compare the relationship of a dog to a table with the relationship between space and time. Yet the situation is quite straightforward: just as a dog is very different from a table, they are nevertheless exactly similar in as much as one is no less a material object than the other; so space and time are quite different, yet in one respect they are strictly similar in that time as well as space represents a continuum of positions. This fact provides us with the key to understanding in what respects time and space must resemble and in what respects they may not. Let me explain in greater detail.

In order to arrive at a correct understanding of why there must be a certain class of properties with respect to which space and time cannot fail to be fully analogous, let us consider a typical aspect of space and time in which the two completely resemble one another. The statement 'Nothing can occupy two places at the same time' is commonly regarded as necessarily true. Is its spatial counterpart also true? I.e., is the statement 'Nothing can occupy two times at the same place' true too? Before we can answer this question we must realise that the first statement is true only if understood in a certain manner but not otherwise. A physical object that is spatially extended will have its different parts occupying different places at the same time. So the first statement is true only if it refers either to things which occupy infinitely small amounts of space or if we refer to all the points occupied by its different parts as one place. Under parallel restrictions the second statement is true too. An infinitesimally short-lived entity cannot occupy two times at the same place. In addition, if we decide to consider the whole period during which an entity is at any single location as one time, then even a temporally extended particular cannot occupy two times at the same place.

All this is obviously true and furthermore is true not only about space and time but about every pair of continua containing some occupant. Suppose there is an occupant that has an extended existence in two continua, X and Y, and any point x in X occupied by the occupant is related to some point y in Y occupied by the same occupant by the relation represented as $y = f(x)$. Suppose that at some point y_1 the continuant occupies many points in X. Then it is correct to say that the occupant 'is at two places in X at the same point in Y.' On the other hand if $y = f(x)$ is such a function that for any particular value of x there is a unique value of y, and for every value of y there is a unique value of x, then the occupant 'cannot be at two places in one continuum while at one place in the other continuum.' Finally, even if corresponding to a particular point occupied in Y many points in X were occupied, and corresponding to a particular point occupied in X many points in Y were occupied, we nonetheless decided to call the whole set of points in X which correspond to a single point in Y as the same place in X, and vice versa, then once more it is true that 'it is impossible for an occupant to be in two places in one continuum while at one point in the other continuum'.

But what has just been said completely reveals the scope of the resemblance between space and time and provides a full explanation why such resemblance has to exist. Space and time are both continua and possess therefore all the properties continua in general possess; events, processes and objects, which are occupants of space and time, also have all the properties which are characteristic of occupants of continua in general. Any statement that is true about an occupant of a pair of continua X and Y (where X and Y have only general continuum properties without any extra properties which would turn them into a specific kind of continuum) must be translatable into a pair of true statements (and not just into a single true statement, as X and Y are interchangeable) specifically referring to space and time. This means that if there is a true statement asserting that an occupant has a specified relation R to any continuum X which it occupies, of necessity it must have also a specified relation S to any second continuum Y which it also occupies, it must be true of some occupant of space-time that if it has relation R to time then it has relation S to space and also true of some occupant that if it has relation R to space then it has relation S to time.

This can also be put in a slightly different manner: a temporal statement that is true not by virtue of any special time-like property of time, or by virtue of any space-like property of space, but exclusively by virtue of those characteristics of space and time which these have as an inevitable consequence of the fact that each one of them constitutes a continuum, must have a true spatial counterpart, as time and space are continua to equal degrees.

Now whereas space and time possess all the properties continua possess in general, it is also true that time possesses an indeterminate number of extra properties which it does not share with any other continuum, and by virtue of which time is specifically time and not just a continuum. The same is true of space. As long as we restrict ourselves to continuum properties, we are dealing with a common denominator of space and time and, hence, we shall find space strictly resembling time. But as soon as we are considering the time-like properties of time, e.g., properties that are not imposed upon time merely by virtue of the fact that it is a continuum, and as soon as we are considering the space-like properties of space, there is no reason why this similarity should exist. (v. Q&A 4, 6, 11 and 12)

6   How to Determine Whether a True Temporal Statement Necessarily Has a True Spatial Counterpart

It is therefore quite a simple task when faced with a true temporal statement to determine whether or not it will of necessity have a true spatial counterpart. First of all we have to try and reformulate the statement in pure continuum terms, making a universal statement concerning any continuum—in case the temporal statement refers to time alone; or concerning any pair of continua —in case the temporal statement refers to space as well as time.

What are continuum terms? All occupants of continua in general occupy points which may stretch or extend over distances which can be divided into intervals. The extremities or two occupants may coincide, in which case the occupants are congruent. Or, they may have only one of their extremities coincide, in which case they are adjacent; they may also be inside one another or overlap or have no common points. Thus terms such as 'point', 'stretch', 'extension', 'distance', 'interval', 'extremity', 'coincide', 'congruent', 'adjacent', 'overlap' are continuum terms. Typical temporal terms or spatial terms may be translated into these in the following manner: the temporal term 'duration' is equivalent to the continuum term 'extension'; 'to occur at t' is equivalent to 'to be at point t'; 'are simultaneous' is equivalent to 'are at the same point', and the spatial term 'the place of O' is equivalent to 'the set of points occupied by O', and so on.

After we have translated the temporal statement into the equivalent continuum statement we must examine whether the latter is necessarily true of all continua. If it is, then the true temporal statement has of necessity a true spatial counterpart; otherwise, it does not.

For example, the temporal statement 'Nothing can occupy two places at the same time' is true if we call all the points by all the parts of an object the same place. We have seen in the previous section how this statement is to be reformulated in pure continuum terms, and that the resulting statement is of necessity true. It follows necessarily that the spatial counterpart of our temporal statement is true; which we found was indeed the case.

On the other hand, the statement, for example, 'There is order in time' translates into something like 'The various points of occupants in all continua are ordered in a special way', which of course is not true. Whether the various points of occupants in a given continuum are ordered in a special way or not depends on the specific nature of that continuum. The specific space-like features of space being what they are do not require that the occupants of space all be ordered in a special way. The spatial counterpart 'There is order in space' need not be, and indeed is not, true.

Or again, by considering Strawson's example, we find that time is a much more basic constituent of the world than space. There is nothing surprising in this difference. The fact that both time and space are continua reveals nothing about their extra-continuum physical properties, which play the crucial role in determining how essential time and space are in sustaining a viable world. Some physical continua are of very little conceptual importance, and the world would undergo no fundamental changes if they did not exist. The specific space-like characteristics of space, however, are such that they lend space a paramount significance. Yet as Strawson has shown, a no-space world that contained reidentifiable particulars is coherently describable. On the other hand the specific time-like characteristics of time, being what they are, render a no-time world conceptually unviable.

Of course there may be philosophers who will not allow that such a difference between the degree of importance which time and space play in the world exists. But even they would have to agree that this is not because both time and space are continua but because their extra-continuum properties happen to imply what they think they imply. It is quite different from those features in which a difference between space and time is inconceivable, for those are general features of all continua.

II

TEMPORAL BECOMING

1 *The Controversy Concerning the Existence of A-Statements*

In this chapter we shall consider what fairly may be called the profoundest issue in the philosophy of time: the status of temporal becoming. Some philosophers have even regarded it as the profoundest issue in all of philosophy; C. D. Broad, for instance, thought so and kept returning to the problem of temporal becoming throughout his life.

According to a view deeply ingrained in all of us, a view explicitly championed by J. E. M. McTaggart, the NOW is something that moves relative to the series of points that constitute time. Temporal points from the future, together with the events that occur at those points, keep approaching the NOW and after momentarily coinciding with it they recede further and further into the past. The NOW is, of course, not conceived as some sort of an object but rather as the point in time at which any individual who is temporally extended is alive, real or Exists with a capital E. I may be occupying all the points between the year 1900, my date of birth, and 2000, the date of my departure from this world, but only one point along this one-hundred-years chunk-of-time is of paramount importance at any given instance, namely, the point that is alive in the present, the point that exists not in my memory or is anticipated by me, but of which I am immediately aware as existing in the present. (v. Q&A 13)

A typical event, on this view, to begin with is in the distant future; then it becomes situated in the less distant future; it keeps approaching us until it becomes an event occuring in the present. As soon as this happens the event loses its presentness and acquires the property of being in the near past. The degree of its pastness continually increases. Thus, events approach us (by 'us' I mean that temporal part of our temporally extended selves which is subject to our direct awareness) from the distant future, become present and then recede further into the past.

According to Bertrand Russell and his followers this is a completely false picture. No event has the monadic property of being in the future, as such, to begin with. Consequently, it can never shed this property. An event, E_1, may occur later than some other event, E_0, but if this is so at all then it is true forever that E_1 occurs later than E_0. Neither can any event be in the past. E_1 may be earlier than E_2, but once more, if this is so then the fact that E_1 occurs earlier than E_2 is an eternal fact. Indeed, all the temporal properties of events and moments are permanent. E_1 has the unchanging relationship of either before, or after, or simultaneous with every other temporal entity in the universe. Apart from moments and the events that occur at them, there is no extra entity such as the NOW, to which E_1 may have a changing relationship. Also, E_1 is as real at t_1 as E_0 is at t_0 and E_2 at t_2; that is, all events are equally real and alive at the times at which they occur and not at others, and they do not come momentarily to life as they are embraced by the NOW. (v. Q&A 14)

The opponents of Russell regard his temporal universe, into which no transient properties are admitted at all, as essentially impoverished, whereas Russellians hold that their opponents admit into their universe nonexistent properties. The fascinating thing about this controversy is that although it is by no means about some remote aspect of the world—for on the contrary it concerns a most immediate and constantly encountered feature of the empirical universe—it nevertheless cannot be resolved within the scope of ordinary observation or of scientific experimentation. Only through philosophical analysis does there seem to be any hope of making some progress toward the resolution of this fundamental controversy affecting one of the most ubiquitous aspects of the universe in which we live. A preliminary requirement, of course, is to understand exactly the two views, both of which have been subject to serious misinterpretation.

The controversy concerning temporal relations expresses itself also in argument about what kinds of temporal statements exist. According to McTaggart[1] there are two fundamentally different kinds of temporal statements—A-statements and B-statements. The latter are the more familiar kind, for B-statements, like all statements in general, have permanent truth values. 'E_1 is before E_2' is a typical B-statement, which if true at any time is true at all times, and if false at any time is false at all times. A-statements, on the other hand, are statements whose truth-value is subject to change. 'E_1 is in the future' is an example of an A-statement, as it is

1. J. M. E. McTaggart, *The Nature of Existence*. Michigan, Scholarly Press, 1968, Volume II, Book 5, Chapter 33.

true if asserted at any time which is earlier than the occurence of E_1 but false if asserted at any other time.

McTaggart considers it essential that there be A-statements, for in their absence there is no possibility for change, and time would not be real if it did not permit change. But change occurs only when a fact that at one time has obtained ceases to obtain at another; or, to put it differently, when a given statement that was true at one time becomes false at another, or vice versa. Russell tried to argue that A-statements may be dispensed with, because changes may be expressed with the aid of B-statements alone as, for example, in the case of a poker that is hot at t_1 but cold at t_2 and thus undergoes a change which manifests itself in the fact that 'The poker is hot at t_1' is true, whereas 'The poker is hot at t_2' is false. To this McTaggart objected that no genuine change in the properties of the poker has been expressed with the aid of these sentences because the first statement is true and never ceases to be true, whereas the second statement is eternally false. In other words, it has been a fact and will always be a fact that at t_1 the poker is hot and, similarly, it is an unchanging fact that at t_2 the poker is not hot. Only the truth-value of the A-statement 'The poker is hot now' really undergoes a change, for the statement is true when asserted at t_1 but false when asserted at t_2.

According to McTaggart, A-statements are those that refer exclusively to temporal properties of events or moments, and in no other domain do we encounter any such peculiar statements. For example, 'O is here' does not have the feature characteristic of A-statements that it changes its truth-value. At first look this may not be clear, for it may seem that it is true when asserted at the same place where O is but false when asserted elsewhere. This however is not really so. I am fairly certain that McTaggart would accept the analysis according to which when asserted at two different places 'O is here' amounts to two different assertions. The correct analysis of 'O is here' is 'O is at the place where I am', so when I am at p_1 the proposition in effect asserts that O is at p_1; but when I am at p_2 then through the same words I assert that O is at p_2. Thus, if O is in fact at p_1 and so am I, and I utter 'O is here', then I make a true assertion which remains always true. If at p_2 I again utter 'O is here', then I make a false assertion, but one which is different from the assertion made at p_1; for now I am asserting in effect that O is at p_2, and this is false and was false in the first place.

According to Russell[2] there are no A-statements. All statements have permanent truth-values. To a sentence such as 'E_1 is in the future' Russell applies basically the same kind of analysis as

2. Bertrand Russell, *Principles of Mathematics*. New York, Norton, Section 442.

everybody does to a sentence such as 'O is here', namely, that when uttered at different times it expresses a different proposition.[3] One variation of this kind of analysis is due to Reichenback and is also embraced by several other philosophers, among them J. J. C. Smart, according to which 'E_1 is in the future' is reduced to the B-statement 'E_1 is after the event of the utterance of this token', where 'this token' refers to the sentence-token just being uttered. Consequently, when this sentence is uttered on two different occasions, once before E_1 and the second time after E_1, the first time it is asserted the proposition is true and is unalterably so. The second time the proposition is asserted, it is a different one, because unlike the first proposition, which claimed that E_1 is later than the first token, it claims that E_1 is later than the second token. The second proposition is now and has always been *false*.[3] (v. Q&A 15–22)

2   *The Definition of an A-Statement*

It is useful to consider briefly Gale's attempt to give an accurate definition of an A-statement:

> Any statement which is not necessarily true (false) is an A-statement if, and only if, it is made through the use of a sentence for which it is possible that it is now used to make a true (false) statement and some past or future use of it makes a false (true) statement even if both statements refer to the same things and the same *places*.[4]

Disregarding for the moment some of the details of this definition, we should note its most significant feature, which is that an A-statement is one that is expressed by a sentence which on different occasions may express statements with different truth-values. According to Gale, it is only the sentence-type that remains the same from one occasion to the other; the statements that are made are different. Thus, even according to McTaggart it is not the case that the self-same statement may have different truth-values on different occasions.

Gale is, however, unquestionably wrong in this. If he were right, it would be unexplainable why McTaggart thought that once we have A-statements we have secured genuine change, for after all, the true statement that we make through 'E_1 is in the future', uttered earlier than E_1, remains unalterably true, and the statement

3. Russell's analysis is eliminative and not reductive, as A-statements are eliminated and we are left with B-statements only.

4. Richard Gale, *The Language of Time*. New York, International, 1968, p. 49.

that is expressed by 'E₁ is in the future' uttered after E₁ that is false, is a different statement. Surely the reason we have change when we have A-statement is because what is said of E₁, namely, that it is in the future, is at one time true, and then the self-same assertion changes truth-value and becomes false. Thus, when the utterance 'E₁ is in the future' is made twice, once before the occurrence of E₁ and once after the occurrence of E₁, we have the very same statement with different truth-values at the respective times. Besides, McTaggart's own writing leaves no doubt concerning this matter, for he explicitly says that the essence of change is embodied by the fact that the self-same proposition changes from true to false, or vice versa:

> It follows from what we have said that there can be no change unless some propositions are sometimes true and sometimes false. This is the case of propositions which deal with the place of anything in the A-series—"the battle of Waterloo is in the past", "it is now raining". But it is not the case with any other proposition.

Interestingly, this passage, which Gale ignores, is reprinted in his own *The Philosophy of Time* (Garden City, 1967) p. 93.

Thus, contrary to Gale, A-statements are statements which themselves undergo changes in truth-value. In view of this fact it becomes evident also that the last phrase in Gale's definition is superfluous. As he himself explains he has added the phrase 'even if both statements refer to the same things and the same places' to exclude, first of all, a statement such as 'I am Richard Gale'. Without a special proviso this statement satisfies his definition of an A-statement, as it is logically possible that two non-simultaneous uses of the sentence expressing it make statements differing in truth-value. This would happen—as he points out—if one of these statements was made by him and the other by someone else. Another example of a kind of statement that he needs to exclude is 'O is here'. But of course on the correct understanding of the nature of A-statements we need make no special provisions to exclude these sentences. The statement expressed by 'I am Richard Gale' in any case does not qualify as an A-statement, because when the sentence is uttered by two different people it makes different statements. Similarly, as we have already said, when uttered at different places 'O is here' makes different statements. It is only a sentence such as 'E₁ is in the future' which, no matter when it is uttered, expresses the same statement—a statement which may on one occasion be true and on another false. (v. Q&A 22, 23)

3 *Some Unsuccessful Attacks on Russell's Position*

Many people intuitively feel dissatisfied with the Russellian analysis of temporal statements. By eliminating A-statements, they feel, Russell is left with an essentially impoverished notion of time. It seems extremely difficult however to give a correct expression to one's dissatisfaction. Gale makes several valiant attempts to do so but does not seem to succeed very well. He considers among others that particular Russellian suggestion according to which the correct analysis of 'S is now Φ' is the statement 'S's being Φ *is* simultaneous with theta' (*is* is a tenseless copula), where 'theta' is a metalinguistic proper name for the occurrence of the tensed sentence token in the analysandum. Gale claims that this analysis must be wrong because:

> The B-statements in the analysans of these two analyses do not entail the A-statements in the analysandum; that S's being Φ *is* simultaneous with theta (the occurrence of a token 'S is now') does not entail that S is now Φ. These B-statements describe a B-relation between S's being Φ and a certain token event without entailing that either of these events is now present (past, future). That they do not convey or entail information about the A-determination of an event can be seen by the fact that whenever someone uses the sentence 'S's being Φ *is* simultaneous with theta (the occurence of a token of "S is now Φ")', he has not forestalled the question whether S's being Φ (or the occurrence of theta) is now present (past, future).[5]

Until now we have been speaking of A-statements, but in the passage just quoted we find the term 'A-determination'. A-determination stands for the peculiar property that may be predicated of events and with respect to which it may change. For example, a given event may be liked by John at t_1 but ceased to be liked by him by t_2; yet it would be wrong to think that the event in question has undergone a change with respect to the property of being liked by John between the times t_1 and t_2. For strictly speaking, it is the property of being liked by John at t_1 which the event has and which it never ceases to have; whereas it lacks permanently the property of being liked by John at t_2. On the other hand, a property such as 'being in the future', as such, is a property in itself, and with respect to this an event may undergo changes according to McTaggart. (v. Q&A 23)

---

5. *Ibid.,* p. 55.

What Gale seems to have overlooked is that the whole point of the Russellian analysis is to show that there is no such thing as A-determination, and all temporal properties are, strictly speaking, permanent properties. Hence, there is no room for complaining that the analysans put forward by Russellians convey or entail no information about the A-determination of S's being Φ, for that event, just like any other event, has no A-determination. Events can have only such unchanging relations as being before, after or simultaneous with other fixed events or moments. There do not exist such extra and variable properties as being before, after or simultaneous with the 'present'. When we say of an event that it is occurring 'now', according to Russell we merely assign to it the property of being simultaneous with some other fixed event. For saying that an event is occurring now is no more than an abbreviated way of saying that the event in question is simultaneous with a given token. On the Russellian view 'S is now Φ' is no more or less than 'S's being Φ *is* simultaneous with theta', and therefore the latter sentence fully expresses all that the first sentence expresses. Only on McTaggart's view does it make sense to ask, after it has been said that S's being Φ is simultaneous with theta, 'But does S's being Φ occur in the present?' which is in effect to ask whether S's being Φ is simultaneous with the shifting present. On the Russellian view there is no such thing as the 'shifting present', and after we have affirmed that S's being Φ has the permanent property of being simultaneous with theta there is no room further to inquire about the variable properties of the same event. There simply are no such properties.

In another attempt to show the inadequacy of the Russellian analysis Gale says:

> . . . that X is now present, unlike the statement that X *is* simultaneous with this token does not entail that there is a token and moreover one which is simultaneous with X. The statement that X is now present seems to have different truth conditions from the statement X *is* simultaneous with this token; for that a token occur simultaneously with X is a truth-condition of the latter but not of the former statement. The statement that X is present although no token occurs, unlike the statement that X *is* simultaneous with this token although no token occurs does not appear to be a contradiction.[6]

Once more what Gale says is true for McTaggart. According to him

6. *Ibid.*, p. 207.

an event simply is or is not in the present, irrespective of any token being uttered. But not so on the Russellian view, according to which being present is a diadic relationship between an event and a token referring to it, and when there is no such token, there is nothing with respect to which the event in question may have this relation. Russell maintains that time is essentially similar to space. Just as it would be wrong to insist that 'O is here' may be true whether I am or am not at the same place as O, so it cannot be said that 'X is present' may be true when X is not simultaneous with any utterance.

We also may mention here an apparent minor flaw in the Russellian view, which however may easily be ironed out. When two different people say at the same time that X is now, common sense tells us that they are making the same assertion. According to Russell, however, each one is claiming that X is simultaneous with his utterance, thus relating X to a different event and thereby making a different assertion. Once more, however, it is useful to compare the temporal and spatial situations. When two different people who are at the same place say that O is here, they may be viewed as expressing the same statement, because both of them claim that O is at the same place where they are; that is, they assert the co-spatiality of O and the same point in space. Similarly, 'X is present' can be taken to mean 'X is simultaneous with the time at which the utterance of this token is occurring', in which case both are asserting the co-temporality of X and the same point in time. (v. Q&A 27, 28)

4   *The Advantages and Disadvantages of the Two Positions on the Nature of Time*

The strongest motive, however, for preferring McTaggart's view to Russell's is the deeply entrenched impression, shared by all people, of the transiency of time and the generally held belief that time is moving. According to Russell there is no room for any transiency, as all temporal relations between events themselves and events and moments are permanent, and no temporal particular changes its fixed position in the temporal series of moments. According to McTaggart, however, it is possible to look upon the 'NOW' as a particular which shifts its position relative to the series of events in the direction of the future. This movement is manifested by the fact that at one stage it is a fact that E_1 is in the future, which means that E_1 is at a point in time which is later than the time at which the 'NOW' is situated. Yet at another stage this ceases to be a fact, and the 'NOW' reaches the same position in time at which E_1 is situated; and the two are simultaneous when, of course, it becomes true that E_1 is in the present.

Now whereas nobody denies that a deeply felt impression that time indeed flows relative to the present is a part of our mental makeup, many philosophers have already cited very strong reasons why this impression must be mistaken. After all, if there really was a relative movement between the 'NOW' and the series of moments, it would make sense to ask how fast this movement took place. A moment's reflection, however, reveals that it is not because we lack this or that information that prevents us from providing an answer to this question, but that it is in principle impossible to measure the speed of this movement, which therefore makes it necessary to deem it nonexistent.

A second and even stronger argument consists in pointing out that a movement always essentially involves two series, so that points in one may be correlated to points in the other. For example, when a car is moving along the road this motion is embodied in the fact that one position of the car in the series of spatial points corresponds to a given point in the series of moments, while a second position of the car in the same series of spatial points corresponds to another point in the series of temporal positions. But how could the movement of the 'NOW' along the series of moments be realized? What other series is there in which two different points correspond to any two positions the 'NOW' occupies along the time series?

Another famous objection to the belief that time flows relative to the present is due to Broad. When a car reaches a given point in space, that is one event; when it reaches another point, that is another event. It is these kinds of events that form the elements of moments that constitute our time series. When the 'NOW' reaches a given point in this series of moments, that must also be some kind of an event, but one that surely cannot be a member of the very set which constitutes that moment. Thus, unless we are prepared to introduce an additional meta-series of moments which are made up of these events, we must deny the reality of these events and resign ourselves to the fact that the 'NOW' hitting moments in time is not something that really occurs.

It should be mentioned that Broad notes that the difficulties attending the notion of temporal becoming could be overcome if we were prepared to postulate a higher-order series of moments. We may see at once that this is so. The movement of the NOW in the standard series of time may be explicated by explaining that the NOW is at t_1 in the ordinary series when it is at T_1 in the super-series, and at t_2 in the ordinary series when it is at T_2 in the super-series. We may even assign a value to the speed of the NOW: it moves from t_1 to to t_2 at the average speed of

$$\frac{t_1 - t_2}{T_1 - T_2}.$$

As to the third objection, the event of the NOW reaching t_1 may if we like be looked on as taking place in super-time. Naturally, we should be very reluctant to postulate a whole new series of temporal points. But Broad warns us that the difficulty is far more serious than that. The introduction of one extra temporal series would not solve our difficulties. If we could not make sense of the first series without postulating a moving NOW, which in turn requires that we postulate a second temporal series, then we shall inevitably find it conceptually necessary to postulate a moving NOW for this new series as well which, in turn, will commit us to a third series, and so on.

But there Broad moves far too quickly. Not only does he not bother to show that the regress is vicious, he fails also to show that there is indeed any regress at all. It is by no means clear that if we wanted to endow the new series with a moving NOW we would have to postulate yet another series. For just as the second series could be instrumental in helping to make sense of temporal becoming in the first series, in the same manner the first series could serve as the extra series through the use of which temporal becoming in the second series makes sense. For example, it might be said that the NOW in the second series is at T'_1 when it is at t'_1 in the first series, and at T'_2 when it is t'_2, leading to the claim that the average speed of the NOW in the second series from

$$T'_1 \text{ to } T'_2 \text{ is } \frac{T'_1 - T'_2}{t'_1 - t'_2}.$$

In order, therefore, to refute McTaggart conclusively three things would have to be done, things that to my knowledge have so far not been attempted by anyone. First, it would have to be shown that McTaggart cannot make sense of the changes going on in or-dinary time unless he postulates a meta-time of equal richness. That is, a meta-time that admitted B-relations only would not be capable of performing its required function. Secondly, it would have to be shown that in order to explicate the movement of the NOW in meta-time, we could not employ standard time in the same manner we employed meta-time to explicate the movement of the NOW in standard time, and therefore we would be forced to introduce a third temporal sense. Lastly, it would have to be shown why the regress thus resulting would have to be regarded as vicious.

If these three things were accomplished, formidable obstacles would rise against giving a coherent account of the transient aspect of time. In spite of this possibility many would remain strongly reluctant to accept the idea that 'X is now' is more than an abbreviation for 'X *is* simultaneous with this token'. Such analysis impoverishes time greatly because it renders all moments equal, as every single moment is equally simultaneous with all the tokens that are uttered at that moment. But all moments in time are by no means equal; there is always a privileged moment. Suppose t_1 is in the present time. Then surely it is not merely the case that t_1 is simultaneous with a certain set of utterances, and the only manner in which t_1 differs from any other time, such as t_2, is that the latter is simultaneous with a different set of events. There is a most significant feature that sets t_1 apart from any other time. All the utterances made at t_1, as well as all other events occurring then, are in the predominant position of having the capacity of being directly presented to that part of one's awareness that is being lived. Along a human consciousness—which may stretch out over several decades—not each point is like every other point. In fact, there is one particular point that is real and alive, while every other point exists only in one's memory or in one's anticipation. All events that are simultaneous with that point in one's consciousness which is being experienced are privileged events, in that they are occuring now.

This move, however, can be defeated. The Russellian will ask which particular moment is privileged in the sense just described at any given time. The only possible answer to this question is that at time t_1, time t_1 is so privileged. But then he too is ready to admit the trivial truism that *at* time t_1, time t_1 is privileged, as is every point of time at its own occurrence. By voicing this truth we are actually claiming that ultimately all points in time are similar to one another.

To conclude this section we may consider another valiant attempt to show that Russell's impoverished account is incapable of accommodating certain obvious features of the universe. Richard Taylor in the last section of his work *Metaphysics*,[7] entitled 'The attempt to expurgate pure becoming', considers four statements which seem impossible to reformulate in terms of B-statements (he himself does not use this terminology). One of these is 'Y is receding even further into the past'. Among other things, he says:

> We cannot just say that Y is *earlier* or *anterior* to some time
> for instance, for this is true of all time whatever, including

7. Englewood Cliffs, New Jersey, Prentice-Hall, 1965, pp. 80-83.

those which are not receding into the past—namely of all future times. We must identify the time to which Y is anterior, either as being the present time or some time itself anterior to the present, and the hopelessness of this is quite obvious (p. 82).

But surely one could suggest a partial translation of Taylor's statement which would go something like 'Y is earlier than this token and even earlier than any token later than this token'—which of course indicates clearly that Y is a past event. Thus, what a Russellian would reply is that part of Taylor's statement can be translated into a B-statement, which will reflect the sense of the original statement, and part of it cannot. But the part that cannot be translated does not truly make sense; and it is rather an advantage that we have adopted a language in which certain senseless sentences cannot be formulated. Thus, the Russellian will freely admit that the original statement said more than merely 'Y is earlier than this token, etc.'; it also suggests that Y is *engaged* in some sort of *activity* of receding into the past, none of which is captured in the translation. But this is just as well. For while recognizing how strongly such notions are fixed in our minds, Russell would insist that they have to be expurgated. The suggested B-statement captures all that is factual in Taylor's statement.

5   *Our Different Attitudes Toward the Past and the Future*

And yet there seem to exist some inescapable facts which may be construed as evidence, not entirely conclusive perhaps, but strongly indicative that the temporal universe does have the richness attributed to it by McTaggart, and that the more austere Russellian view concerning what temporal relations exist is inadequate. The evidence I shall be concerned with here consists in our very different attitudes that prevail toward the future and the past. The existence of these differences is universally acknowledged and is shared by Russellians no less than by others. Nobody denounces these differences in attitudes as irrational; nobody advocates that our attitudes are to be reformed in the light of a clearheaded analysis of temporal relations. But there is a strong case for claiming that these differences are justified only if time also has a transient aspect, and moments do partake in a movement the direction of which is specifically from the future toward the past. On the Russellian view, which does not permit any changes and according to which all temporal relations are permanently fixed, it is very hard to justify such differences in attitude.

Consider the difference in our attitude toward a very unpleasant experience, such as a painful operation, which was performed on our body in the past, and our attitude toward the same kind of event which we know we are going to experience at some given time in the future. In the first case thinking about the harrowing experience is accompanied by a feeling of relief; in the second case contemplating the experience arouses in us a feeling of anxiety and dread. Now why the relief in the first case? Obviously, because the highly disagreeable experience is 'over', that is, it is receding from us and we are escaping from it rather than still experiencing it or moving toward it. On the other hand the feeling of dread in the second case is explained by the fact that the agonizing experience is seen to be approaching us and is known to be about to overtake us.

On the Russellian view, however, there is no room for such talk, because no events are receding from us, and none is approaching us. Distances between all temporal particulars remain fixed permanently. Consider for instance our state of mind at t_2, which lies between the two times t_1 and t_3, at both of which a painful event occurs. Neither of these events is being experienced at t_2, so why are we at that time so concerned about the event at t_3 in the future? The explanation is that it is thought to be threatening because it is *going* to be experienced by us. But this feeling of being threatened seems to make sense only if 'going to be experienced' is understood not merely as claimed in the last section 'it is experienced at a time which is later than t_2', but if it is taken to have the stronger connotation 'it is shifting its position toward us'. But it is only according to McTaggart that it is legitimate to think of events as engaged in the process of moving toward or away from the present. As we have already said, according to Russell time is essentially like space, in which all relations are fixed. If it is given that I have a painful experience at a spot that is one mile to my left and also one at a spot that is one mile to my right, from this we cannot derive that there must be a difference in my attitude toward these two experiences. The spatial relations of these two relations are symmetrical with respect to my position, and knowing them does not by itself warrant that I should be concerned more by the one than by the other. Similarly, when an unpleasant experience occurs at a given temporal distance from the time at which this token occurs, why should it matter in which direction this experience lies?

Another question that we may raise is, why should our attitude to the event of our birth differ from our attitude to the event of our death? Both events represent a dividing point between a period of existence and nonexistence. Why does it matter so crucially that in one case the period of nonexistence lies in the direction of before, to the point of division, and in the other in the direction of after?

Russellians perhaps might want to reply that the difference lies in the fact that in the first instance one begins to exist; in the other, one ceases to exist. But to them the time at which one begins to exist means no more than the time later than which one does exist, earlier than which one does not. And the time at which one ceases to exist is time later than which one does not exist, earlier than which one does. The question then once more returns, but why is it so crucially important on which side of the division point the periods of existence and of nonexistence lie? On McTaggart's view, of course, matters seem to be self-explanatory. To begin to exist means to move away from a state of nonexistence into a state of existence; and given that it is desirable to exist, births are occasions for joy. To cease to exist, on the other hand, is to move away from a state of existence into a state of nonexistence, which is something to bemoan.

An entirely similar point emerges from the difference between our attitudes toward pleasant events which are known to have occurred in the past and those which are known to be occurring in the future. Pleasant experiences of the past are recalled with nostalgia, and we regret their passing; that is, we are sorry that they are getting further away from the 'NOW', which is the point in time at which events occurring at that time are real to our experience. On the other hand pleasant experiences are being looked forward to with joy, for they are approaching the 'NOW' and are about to overtake us.

Philosophers such as Donald Williams and J. J. C. Smart have argued eloquently that it is completely wrongheaded to speak of the 'river of time' or of time as 'flowing', 'marching on', and the like, for moments and events do not partake in any movement. The relative positions of all temporal particulars with respect to one another are fixed forever. Consequently, at a point in time t_2, which is later than t_1 and earlier than t_3, an individual is entirely symmetrically related to t_1 and t_3 in the sense that he is no more moving toward the one than toward the other. There is nothing in the nature of time that could explain a greater concern at t_2 with t_3 than with t_1. Yet nobody has deplored our attitudes to different sorts of events of the future and the past. Nobody has advocated that our states of mind ought to be the same when contemplating our death as when contemplating our birth; or that we should be no more pleased by the pleasant events the future holds for us than by those that occurred in the past; nor should we dread the misfortunes of the future more than those of the past. (v. Q&A 29, 30, 31)

6   *The Unnatural Attitude Required by Russell*

Perhaps it may seem to some Russellians that their greater concern with the future than with the past could be justified simply by the undisputed fact that causes precede their effects. Russellians agree that at t_2 we may affect events occurring later at t_3, but not events occurring earlier at t_1. Hence, fear at t_2 of the disasters of t_3 fulfills the useful function of spurring us into action at t_2, which may prevent those disasters. Also, the thrill with which we anticipate at t_2 the happy events of t_3 plays the role of inducing us to help these events to come about. Concern at t_2 with the happenings of t_1 has no such usefulness.

This defense will surely not do, as we dread no less the calamities of the future which are absolutely unpreventable and look forward no less jubilantly to great pleasures which are sure to materialize without our help. And it seems that we do not find such attitudes irrational.

The only way out of this difficulty would seem to be to deny what has been said in the last section and insist that because time is no more flowing in one direction than in another, a true philosopher ought indeed to face the future with no less equanimity than the past.[8] The enlightened thinker will view events ahead with the same detachment as the ordinary person views past events which have no further repercussions.

This strikes me as a most unlikely solution. Nor does it seem helpful to summon to mind the traditional notion of the philosopher as being calm and unperturbed at all times; to cite the example of Socrates, who was without fear, even of death; to refer to the fortitude of the Stoics, and so on. The special attitude that has been typical of some philosophers, which we feel is appropriate and even admirable, is based on something entirely different. These sages had a different perception of what is of real and lasting value, e. g., life after death, the possession of virtue, and so on; and, being single-mindedly engaged in the pursuance of their higher goals, they became indifferent to the mundane and transient pleasures and pains which excite others.

A simple test that distinguishes between such traditional attitudes and the one that is implied by the Russellian interpretation of time is this: Socrates, the Stoics and other such thinkers allegedly were as composed and calm about the present as they were about the future, yet the Russellian view of time has no implications whatever concerning the required attitude toward what is going on now. We know that it is by no means contrary to human nature to view more or less with indifference one's past and future as well as present fate. All of us have encountered people who, to some degree, emulate the attitudes of these luminaries of the past and—because

8. In conversation Professor Smart has suggested—tentatively—this as a possibility.

they are absorbed in the world of the intellect, the spirit, the soul or morality—remain relatively unperturbed by what has happened, will happen, or is happening to them. The Russellian doctrine of time, on the other hand, permits one to rejoice in pleasures one is experiencing now and also to bewail one's present perplexities. It is only about events of the future that one must not feel any different-ly than about those of the past. I do not believe that many of us have met people exhibiting such strange attitudes. Nor do I believe that, should the Russellian idea of the stillness of time gain wider acceptance, such peculiar partial philosophical equanimity is going to be manifested by many more people.

The last sentence of the last section should therefore read 'Nobody, *or hardly anybody, who has been concerned con-siderably with his present well-being and comforts,* has advocated that our states of mind ought to be the same when contemplating our death as when contemplating our birth, etc.' The sentiments re-quired by the Russellian doctrine of time seem contrary to human nature. They seem contrary to human nature according to their own admission since they have refrained from advocating that in light of their analysis of temporal statements we ought to adopt these sentiments. (v. Q&A 32)

7   *On the Meaningfulness of Certain Temporal Sentences*

Finally, we shall consider an argument of an entirely different nature against Russell, based on the fact that he is obliged to judge devoid of meaning certain expressions which strike most of us as very meaningful. Let us suppose that it is January 1, 1978 $(=t_1)$ and I am looking back on the last ten years, greatly regretting the occurrence of some events during that period which have adversely affected my life. Suppose I do not believe in complete physical determinism (and although I may be mistaken about the nature of the universe, surely I am not incoherent in maintaining that it is in-deterministic), but choose to believe that things could have turned out differently from what they actually have, so that some of the crucial events e_1, e_2, . . . e_n which had unpleasant consequences for me could have failed to occur. In view of this I fervently wish that it was now January 1, 1968 $(=t_0)$, which would amount to my having the advantage of there still being a chance that one or more or e_1, e_2 . . . e_n might not be going to happen. Things being as they are, the history of the world during the interval $t_0 - t_1$ is given to have been in a certain way, and I am not going as far as to ask that this interval be definitely different but merely that we be in a situation in which there is still a real chance of it turning out dif-ferently. My wish is, of course, futile, having no chance of fulfillment. But there is no lack of clarity as to what exactly I am asking for, nor as to why. Anybody familiar with my plight would fully sympathize with me and unfailingly grasp what feature of the universe I should like to be different from what it is: instead of the NOW being at t_1, I should like it to be at t_0.

But a Russellian could not give a coherent account of what I am so fervently wishing. At t_1, the nature of the interval $t_0 - t_1$ is completely fixed, and I do not wish it by any means to be otherwise. At t_0, on the other hand, the nature of the interval $t_0 - t_1$ is open and is so without my wishing it. What feature of the universe is it then that I want to be different from what in fact it happens to be?

I, of course, should like to be ten years younger than I actually am. But the Russellian finds it difficult to make sense of my wish. I am not asking to have been born ten years later than I was actually born. Nor do I wish that the time that elapses between the date of my birth and t_1 be shorter than it is. At t_1 I am of a certain age and I do not wish to be a single day younger at that time. At t_0 I am ten years younger than I am at t_1, but this is so anyhow and I need not wish for it. So *when* is it that I am ten years older than I would like to be? Not at t_1 nor at t_0. According to the non-Russellian, of course, it is *now* when I wish to be ten years younger. I should like a change in my age to come about not through a change in my age at any given point in time but by having the NOW situated at t_0 rather than at t_1, when I am ten years older. Russell, however, can not account for this. I cannot ask the NOW to be somewhere else when it is nowhere, for it does not exist at all. The course that seems open for a Russellian is to claim that indeed my wish makes no sense, for the NOW is not a thing that may shift its position. It is hard to reconcile oneself to the fact that such a seemingly well-founded and straightforward wish after all makes no sense whatever.

We have thus briefly reviewed some of the difficulties which stand in the way of understanding a fundamental feature of time. These difficulties call for a great deal more work on the status of becoming and the transient aspect, if any, of time. If it should turn out that the Russellian view is indeed untenable, then it may have to be conceded that time does have a dynamic aspect, even though the movement of the NOW may have to be understood as a very special case of movement, perhaps quite unsimilar to movement in general—one which may not require the presence of factors normally required for movement. (v. Q&A 33, 34)

III

McTAGGART AND
HIS COMMENTATORS
ON THE UNREALITY OF TIME

1 *Preliminaries to McTaggart's Argument*

I may assert confidently that no other thesis advanced in the twentieth century in the philosophy of time has attracted as much attention as McTaggart's thesis that time is unreal. Most discussions of it endeavor to refute the thesis. I shall not attempt to explain what McTaggart might mean by his conclusion that time is unreal. It is after all a most puzzling contention. Everything happens in time; thus, if time is unreal it would be reasonable to infer that every physical particular is unreal too. But it is rather odd to declare the whole of reality as unreal. By the principle of vacuous contrast, if nothing is real, then being unreal no longer carries with it a very ominous connotation.

However, the claim that time is unreal is not an intrinsic part of McTaggart's argument. His argument is designed principally to show that the notion of an A-series gives rise to a contradiction. If his argument is correct, it implies only that the A-series is unreal. The most natural conclusion to draw from this would be that time should be thought of as consisting of the B-series alone. This has been the view of many philosophers, as we saw in the last chapter. But if one believes, as McTaggart does, that the B-series alone cannot give us time, then, of course, one is tempted to conclude that the notion of time itself is involved in a contradiction. Nevertheless, rather than draw from this the strange inference that time is unreal, one might claim tentatively that a difficulty has been uncovered which will have to be resolved before one can arrive at a correct understanding of the nature of time.

Before attempting to grapple with the main argument, I shall consider a preliminary claim McTaggart makes prior to his presentation of the alleged contradiction involved in the notion of time. McTaggart begins by asserting that he is going to show that the

41

A-series cannot be real and that this fact destroys the reality of time, because the B-series cannot stand by itself. To establish this last claim he resorts to two arguments. One argument points out that anyone who understands what A-properties are can understand B-properties, but not vice versa. This shows that B-properties are derivative, whereas A-properties are ultimate for they are undefinable in terms of B-properties.

Thus, the first thing I shall do is to examine the claim that B-properties may be defined in terms of A-properties, or B-statements in terms of A-statements. However, I have to preface this with the remark that the whole matter does not have the importance McTaggart attributes to it. B-properties may be definable in terms of A-properties, yet if there are independent reasons why we must reject A-properties that does not imply that we must also reject B-properties. These now might have to be thought of as undefinable primitive properties. Russell might very well have agreed, for instance, that if there had been such things as A-statements, then B-statements could have been defined in terms of them. However, it so happens that there simply are no true A-statements; but this fact does not prevent the existence of B-properties, which alone are sufficient to give us time.

I shall try to defend McTaggart's claim that B-statements are definable in terms of A-statements, without claiming that anything so dramatic as the destruction of time follows, because a detailed defense of his view leads to a better understanding of the nature of temporal statements.

2   *The Proposed Reduction of B-Statements*

According to McTaggart, for example, the B-expression 'earlier than' may be defined in terms of A-expressions such as 'past,' 'present,' and 'future' in the following manner:

The term P is earlier than the term Q if it is ever past while Q is present, or present while Q is future.[1]

Richard Gale, in his book *The Language of Time,* to which we already have had occasion to refer, has rendered great service by probing into various questions concerning the metaphysical aspects of time. Among these is the question whether indeed B-expressions can be defined in terms of A-expressions. He concludes that McTaggart's theory is false because it involves a circularity:

---

1. J. M. E. McTaggart, *The Nature of Existence,* Cambridge, 1921-7, Vol. 2, p. 271.

> The definiens of the definition contains the statement 'P
> *is* [2] past while Q *is* present' and 'P *is* present while Q *is*
> future' which means the same as respectively 'P *is* past at Q'
> and 'Q *is* future at P'. The predicates '____ *is* past at ____'
> and '____ *is* future at ____' are synonyms for '____ *is* earlier
> than ____' and '____ *is* later than' respectively, in that all
> four are tenseless two-place predicates which, when event
> expressions not containing A-expressions are substituted for
> their blank spaces, express a timelessly true or false state-
> ment about a B-relation between two events, i. e., they make
> B-statements (p. 90).

What is characteristic of an A-statement is that it assigns a temporal predicate to some event or moment and is not timelessly true,
as at one moment the statement may be true whereas at a later or
earlier moment the self-same statement may be false.

> McTaggart is cheating by allowing '*is* past (future) at' to
> count as A-expressions (Gale, p. 91).

In attempting to remedy this situation Gale makes several instructive attempts to give a different definition of the term 'earlier than',
one of which is:

> (II)   P *is* earlier than Q = P is past and Q is present *or* P is
> past and Q is future *or* P is present and Q is future *or* P is
> more past than Q *or* Q is more future than P (p. 92).

He is not certain however about the effectiveness of (II) either. He
says:

> It might be objected, and not without plausibility, that these
> concepts involve a tenseless B-relation, because a statement
> of the form 'P is now more past than Q' must be *analysed*
> into a conjunction of the B-statement 'P *is* earlier than Q'
> and the A-statement 'Q is now past.[3] And if this is so, our

---

[2. *is* denotes the tenseless copula as already pointed out in the previous chapter.]

[3. Gale does not state explicitly what he regards as basic statements and what
statements must be analysed into a conjunction of other statements. But it is obviously most reasonable to assume that he would regard any statement which can
possibly be analysed into components as nonbasic. 'O is now past', for instance, is a
basic statement simply because there is no way in which it could be exhibited as a
conjunction of two statements.]

reduction of B-relations to A-determination by reduction
(II) is circular, since the final two disjuncts of the *analysans*
—the ones using 'more past' and 'more future'—contain
B-statements (pp. 95–6).

In the next section I shall examine more closely Gale's argument
why Definition (II) is not acceptable. I shall render explicit the
presuppositions necessary for Gale's position. In the section after
that I construct an argument to show that we cannot escape the
conclusion that B-statements in fact are definable in terms of a disjunction of A-statements along the lines McTaggart has suggested.
(v. Q&A 35)

3 *The Qualifications of an A-Statement*

I do not believe that anyone would wish to deny that 'P and Q occur in the present' is an A-statement. But if what Gale said in the
last quoted passage were true, it would seem that 'P and Q occur in
the present' should not be allowed to qualify as a genuine A-
statement either. Admittedly, as it stands it does have the
characteristic of an A-statement, in that its truth-value may
undergo changes: given that P and Q occur at t_o, the statement is
true if asserted at t_o but false if asserted at any other time.
However, the statement may be analysed into two components, 'P
is in the present' and 'Q *is* simultaneous with P'; that is, it may be
exhibited as a conjunction of an A-statement and a B-statement.
The fact that this argument would be rejected universally seems to
indicate that we do not care into what components a statement may
be broken. Provided that as it stands it has the characteristics of an
A-statement, it is a genuine A-statement. Consequently, 'P is now
more past than Q' should be regarded as a genuine A-statement,
for, not broken up into its components but left as it stands, it may
be true when asserted at certain times but false when asserted at
others.

In order to meet this objection we may devise the following argument on Gale's behalf: for a statement to qualify as an
A-statement, it is not sufficient that when left in its original form it
may undergo changes in truth-value. We also have to break it into
its components and see what kind of statements these are. Suppose
that when it is broken up into components *p* and *q,* one of them
turns out to be a B-statement; nonetheless, that in itself is not sufficient to bar the original statement from qualifying as an A-
statement. As long as it can *also* be broken up into components *r*
and *s,* where both *r* and *s* are A-statements, the original statement
still qualifies as a genuine A-statement. Only when we can find no

way of breaking it into components, all of which are A-statements, does the original statement, in spite of not having a permanent truth-value, fail to amount to an A-statement.

Thus, in order that a statement should qualify as an A-statement it has to meet two conditions:

(i)   As it stands it may undergo changes in truth-value.

(ii)   There is at least one way in which it may be broken up into components such that all of them are A-statements (that is, all of them are such that they are capable of undergoing changes in truth-value, and when they themselves are broken up into components, it does not result in the creation of any component which is a B-statement). (v. Q&A 36)

I leave it to the judgment of the reader to decide how plausible this argument is. But, if accepted, it does eliminate the difficulty I have posed. Admittedly, 'P and Q are in the present' may be broken up into components in such a way that one of them is a B-statement. However, it also may be broken up into different components, namely, 'P is in the present' and 'Q is in the present', where both components are A-statements. According to the principle just enunciated, 'P and Q are in the present' is not disqualified after all from being an A-statement. On the other hand, 'P is more in the past than Q' may be broken up into 'Q is in the past' and 'P *is* earlier than Q', where the second statement is a B-statement, and we can find no other way of breaking it into components which are all A-statements. This statement therefore fails to qualify as an A-statement. It should be noted that 'P is more in the past than Q' is not merely a B-statement either, as it may have changing truth-values. The statement may be true at any time later than Q but must be false earlier. Thus, Gale is committed to the view that statements referring to the temporal properties of events may fail to be either simply A-statements or B-statements. Perhaps they are to be regarded precisely as compounds of both.

But still it may seem that the following difficulty could be raised: the simpler 'P is in the present' is undoubtedly an A-statement. But it too may be broken up into two statements, namely, 'P is at t' and 't is at present', where the first of these statements is a B-statement. There does not seem to be any way in which 'P is at present' may be broken into components, all of which are A-statements. Consequently, by the principle I have suggested on Gale's behalf, 'P is at present'—which is a paradigmatic A-statement—should fail to qualify as a genuine A-statement.

To this argument, however, one may reply that 'P is at *t*' is not a statement on its own but only a statement form, as the value of '*t*' is not specified, and we are not given what point in time '*t*' stands for. Hence, we cannot be said to have here two fully fledged statements, one of which is a B-statement, because the meaning of 'P is at *t*' does not become determinate until we are told that *t* is in the present. Unless *t* is individuated by a unique event, other than P, occurring at *t*, or by a calendar date and clock time, 'P is at *t*' and '*t* is at present' are not two statements. What they say in effect is (∃t) (P is at *t* and t is at present) which is not merely logically equivalent but amounts to nothing more or less than, and thus has exactly the same sense as, the single A-statement 'P is at present'. (v. Q&A 37)

4   *The Vindication of the Reduction of B-statement*

From what I have said it follows that 'P is past at *t*' and '*t* is in the present' also cannot be looked on as a conjunction of a B-statement and an A-statement, for 'P is past at *t*' remains indeterminate as long as we are not given '*t* is in the present'. It follows therefore that 'P is past at *t* and *t* is in the present' has exactly the same status as 'P is past now'—it is an A-statement.

Now consider the more complex statement:

(*s*)   P is past at *t* and *t* is in the present *and* Q is present at *t* and *t* is in the present.

Surely (*s*) must be an A-statement, as it satisfies both of the two conditions that are sufficient to render any statement an A-statement: (i) The compound statement (*s*) as it stands may undergo changes in truth-value; (ii) each of its components is an A-statement.

It may be noted that (*s*) could also be broken up into different components, e. g.:

P *is* before Q *and* Q is present at *t* and *t* is in the present, in which case the first statement would be a B-statement. But according to the principle suggested on Gale's behalf, a principle that was required for the defence of his position, this does not prevent (*s*) from qualifying as an A-statement.

For convenience's sake we may abbreviate (*s*) into (*u*):

(*u*)   P is past at *t* and Q is present at *t* and *t* is in the present.

We shall also have to agree that as (*u*) is an A-statement, so is

(v)   P is past at *t* and Q is present at *t* and *t* is in the past

as well as

(w)   P is past at *t* and Q is present at *t* and *t* is in the future.

Consequently, the possibility presents itself of defining 'P *is* before Q' in terms of a disjunction of the three statements (*u*), (*v*), and (*w*). In other words we have:

> P *is* before Q = P is past at *t* and Q is present at *t* and *t* is in the present *or* P is past at *t* and Q is present at *t* and *t* is in the past *or* P is past at *t* and Q is present at *t* and *t* is in the future.

The only way in which to fault this demonstration that a B-statement may be defined in terms of a disjunction of A-statements is to deny the principle that in order for all of its components to be A-statements, it is sufficient that there be at least one way in which a complex statement may be broken up to yield A-statements only. In other words, one might attempt to prevent (*u*) from qualifying as an A-statement by pointing out that it may also be broken apart differently, into 'P *is* before Q *and* Q is present at *t* and *t* is in the present', in which case the first component is a B-statement. But this is untenable because then we also would have to disqualify 'P and Q are in the present', which is undoubtedly an A-statement, but one which I have already shown can be broken up into components in such a way that one of the components is a B-statement.

We are therefore compelled to agree that the above definition is legitimate. But that definition may be taken quite plausibly as a reconstruction of McTaggart's claim that we may say that 'the term P is earlier than the term Q, if it is ever past while Q is present'.

5   *An Unusual Interpretation of McTaggart*

Now we come to the core of McTaggart's thesis. His main argument has been reconstructed in several ways, one of which involves an interesting suggestion by M. Dummett[4] on how McTaggart is to be understood. According to Dummett, McTaggart argues that the predicates 'past' 'present' and 'future' are contrary predicates, yet they apply to every event. (It should be noted that from the way in which Dummett proposed to resolve this difficulty, it becomes obvious that the problem according to him is that we assign these

---

4. *Philosophical Review*, 1960.

predicates to the same event at the *same* time.) To remove the inconsistency he introduces second level predicates, of which there are nine:

$$\left\{\begin{array}{c} \text{past} \\ \text{present} \\ \text{future} \end{array}\right\} \text{ in the } \left\{\begin{array}{c} \text{past} \\ \text{present} \\ \text{future} \end{array}\right\}$$

Dummett points out that the predicates which we assign to the same event now are not simply 'past' and 'future' but rather 'past in the future' and 'future in the past', which are not contrary predicates. (Indeed, these predicates may be assigned simultaneously to the same event, for it can truly be said that an event which is taking place at this very moment will be past in the future and has been future in the past). But Dummett goes on to point out that among the second level predicates there are some that are incompatible with others, such as 'past in the past' and 'future in the future'; yet these too may be assigned to all events. This inconsistency is in turn removed by the introduction of third level predicates, of which there are twenty-seven:

$$\left\{\begin{array}{c} \text{past} \\ \text{present} \\ \text{future} \end{array}\right\} \text{ in the } \left\{\begin{array}{c} \text{past} \\ \text{present} \\ \text{future} \end{array}\right\} \text{ in the } \left\{\begin{array}{c} \text{past} \\ \text{present} \\ \text{future} \end{array}\right\}$$

The difficulty is removed by pointing out that the same event is not simply 'past in the past' and 'future in the future' but rather 'past in the past in the future' and 'future in the future in the past'.

But among the third level predicates we also find some that are incompatible with one another and we may yet assign them to the same event at the same time; and these inconsistencies can once more be removed by ascending to yet a higher level of predicates, and so on. We note that every inconsistency we find can be removed, but no matter how many are removed, new inconsistencies can be generated ad infinitum.

It does not seem likely, however, that Dummett's interpretation of McTaggart is correct, because it is involved in too many obvious difficulties. The first and chief difficulty is why McTaggart should have thought it necessary to take even the first step along his regress, when there does not seem to be any problem to begin with. Dummett attempts to explain that the problem starts with the fact that 'past' and 'future' are incompatible predicates, yet we may assign them to the same event at the same time. But surely there is no reason to believe that 'past' and 'future' as such can be assigned to the same event simultaneously. It seems clear that only 'past in the future' and 'future in the past' may be assigned together; that is, only when the term 'past' is qualified by the words ' . . . in the future', and the term 'future' by the words ' . . . in the past', may these two terms—which are no longer incompatible—be assigned to the same event at the same time. But then the answer has been provided before any question could be raised.

One could also add to this thought the following one: if Dummett's problem is real, then one might well ask, why is it that 'red' and 'green', which are incompatible predicates, nonetheless may be assigned the same surface at the same time? That this is so may be illustrated by the fact that one may assert with certainty that at a given time a surface was red all over and assert with equal certainty at the same time that the same surface will be green all over at a future time. Of course we are able to assert that redness and greenness apply to the same surface without difficulty because we are assigning red to the surface in the past and greenness in the future. But if there is no difficulty concerning the predicates 'red' and 'green', as indeed no one has ever suggested there is, there is no difficulty with the predicates 'past' and 'future' either. When we assign pastness and futurity to the same event at the same time, it must be clear that we do so correctly only if we assign the former to apply at a later time than the latter.

But let us suppose for a moment that Dummett was right, and there truly is a difficulty with first order temporal predicates. We shall accept Dummett's claim that these difficulties can be made to disappear with the introduction of second order predicates. But then we discover that we can raise a new problem with some of the second order predicates, which in turn is resolved with the introduction of third order predicates, and so on. In this manner a regress is generated. But is it a vicious regress? Admittedly, if we look at matters pessimistically we may say that no matter how many problems we solve, new ones can always be raised; and there is no end to this phenomenon. But why not look at matters optimistically? After all, it may be said also that in spite of the number of problems we raise, we can resolve each one, for any inconsistency we discover among predicates of level n may be removed with the introduction of predicates of level $n + 1$. As there are these two ways in which to appraise our situation, a compelling reason would have to be offered to support the claim that what really matters is that we can raise a new problem after every solution rather than that for every problem there is a solution, before it can be established that we are facing here an infinite vicious regress.

6   The Standard Interpretation of McTaggart

We shall consider now the more widely accepted method of reconstructing McTaggart's argument, according to which 'past' and 'future', as such, apply to the same term, without qualifying 'past' by 'in the future' and 'future' by 'in the past'. The statements 'E is in the future', 'E is in the present' and 'E is in the past' are incompatible statements yet as they stand all are true of E. To remove the contradiction it should be pointed out that these statements, which are indeed incompatible, may all be truly asserted only if asserted at different times. 'E is in the future' may be true if asserted at M_1, whereas 'E is in the present' is true if asserted at a later time M_2, and 'E is in the past' is true if asserted at a yet later time M_3. This neatly removes the problem we found concerning the incompatible statements which are true of E. But what about statements we may assert about M_1, for example? It seems that three incompatible statements, similar to the statements made about E, can be formulated, i. e., 'M_1 is in the future', etc. The answer of course is that these difficulties, too, can be eliminated in the way in which we eliminated the difficulties concerning E: the incompatible statements are true only if asserted at different times, that is, 'M_1 is in the future' is true when asserted at M_4 and so on. But as soon as we remove the difficulty concerning M_1, we raise difficulties about M_4, and so on, ad infinitum. No matter how many difficulties we solve, new ones can always be generated. This involves us in an infinite regress, as a result of which we must abandon the view that properties such as being in the future, in the present or in the past are real properties of events or dates. But if such basic temporal properties do not really exist, then it would seem that to McTaggart time is altogether not real.

The above argument however sounds very unconvincing. In order to bring out clearly its weaknesses, let us consider three simple and seemingly very powerful objections to it:

OBJ. 1:   C. D. Broad in Volume II of his book *Examination of McTaggart's Philosophy,* Cambridge, 1938, p. 313, says:

> I cannot myself see that there is any contradiction to be avoided. When it is said that pastness, presentness, and futurity are incompatible predicates, this is true only in the sense that no one term could have them *simultaneously* or *timelessly.* Now no term ever appears to have any of them

timelessly and no term ever appears to have any two of them simultaneously.

What appears to be the case is that certain terms have them *successively.* Thus there is nothing in the temporal appearances to suggest that there is a contradiction to be avoided.

Broad's objection is very similar to the first objection we raised to Dummett's version of McTaggart's argument. It is difficult to believe that McTaggart would wish to deny that statements 'P is (now) Q' and 'P is (now) not Q' are quite compatible, unless Q stands for the kind of property which when possessed by something at some time must be possessed by it at all times, or both statements are asserted at the same moment in time. Consequently, in order to claim that the temporal predicates 'future' 'present' and 'past' harbor a contradiction, it would have to be shown that these predicates apply to events timelessly and this McTaggart has not done. The only other way in which to create a problem would be to show that statements such as 'E is in the future' and 'E is in the past' may be truly asserted simultaneously, a feat that McTaggart has not attempted. Thus, no difficulty exists to begin with; one is not required to take any steps to avoid a contradiction; and the regress fails to arise.

OBJ. 2:   But let us ignore for a moment Broad's objection and suppose there is a difficulty concerning 'E is in the future' and 'E is in the past', and that this difficulty is not removed until it is pointed out that these statements are truly asserted only at different times. We may then raise exactly the same objection to McTaggart's regress in its present version as the one we raised against Dummett's version. Once more we can look at the regress and see that after every solution offered we can generate a new problem, but at the same time we also see that no matter how many problems we raise, we can always come up with a solution. Before it could be definitely claimed that we are faced with a vicious regress, it would have to be shown that one manner of looking at matters was more valid than another. But no argument has been provided to show that the really important feature of our regress is that every solution can at once be followed by the raising of a new problem rather than that every problem can immediately be provided with a solution.

Arthur Prior, in his *Past, Present and Future,* Oxford, 1967, on pages 5–6 raises what seems to be essentially the objection I have just raised:

Even if we are somehow compelled to move forward in this
way [i. e. taking further and further steps along the regress]
we only get contradictions half the time, and it is not obvious why we should regard these rather than their running
mates as the correct stopping-points. But why do we have to
make the wrong moves in any case? At least after the first
few times, when we've seen the trouble it gets us into, why
not pass to the corrected version immediately?

OBJ. 3:   McTaggart concludes that time is unreal, but apparently
it has never occurred to him to claim in similar fashion that space
also may not be real. However Dummett, in the paper quoted,
raises the question against his own version of McTaggart that the
same objection regarding the reality of time can be raised also
against the reality of space. The version of McTaggart under review
can be similarly attacked as follows:

'Here' and 'there' are contrary terms, yet 'O is here' and 'O is
there' can be asserted simultaneously of the same object O. If it is
replied that 'O is here' is truly asserted at L_1, whereas 'O is there' is
truly asserted at a different point in space L_2, then once more we
raise the same question with respect to L_1, to which the contrary
predicates 'here' and 'there' may be applied simultaneously, and so
on, ad infinitum.

If McTaggart did not think that a problem actually exists here,
why did he think there was such a problem concerning the temporal
predicates 'future,' 'present' and 'past'?

7   *The Solution to the Problems with the Standard Interpretation*

On a more careful look, however, this version of McTaggart's
argument, according to which temporal statements asserted at different moments are incompatible, can be defended against all these
objections. It might be a difficult and unrewarding task to ascertain
to what degree my version can be squeezed into the actual words
McTaggart uses, but what really matters is that the basic structure
of my argument seems to be identical with that of McTaggart. This
structure, I hope to show, is sound. We begin with the following
three propositions:

'E is in the future' is true when asserted at M_1 . . . (a)

'E is in the present' is true when asserted at M_2 . . (b)

'E is in the past' is true when asserted at M_3 . . . . (c)

On the surface of it (a), (b) and (c) may seem quite innocuous; supposing E to happen at M_2 which is after M_1 and before M_3, it seems that all three propositions are true, and thus we face no difficulties. Looked at from a certain point of view, however, these propositions may seem troublesome. Consider the following statements:

'P is Q' is true when asserted at M_1 . . . . (d)

'P is not Q' is true when asserted at M_2 . . . . . . (e)

If P is a temporally extended particular such as a physical object, and Q is some property, then (d) and (e) do not necessarily contradict one another. It is possible for a temporally extended particular to have a given property at a certain moment M_1, subsequently to shed that property, and to fail to have it by moment M_2. But if P stands for a particular which is not temporally extended, such as an instantaneous event, then no matter what property Q stands for, it is impossible for both (d) and (e) to be true. Because E is not extended, that is, there are no two points in time occupied by it, at one of which it could have Q and at the other it could lack it, E simply does not have the scope to accommodate the incompatible properties Q and not Q. In light of this, difficulties appear in (a), (b) and (c). (a) for example assigns a certain property to E, namely, futurity; and (b) denies this very same property to it. How can they both be true? (v. Q&A 38 and 39)

In order to reply to this question note that if P, which occurs at M_1, possesses a property Q (where Q stands for loudness, shrillness, brightness, etc.) then at M_2 or at any other moment P is still qualified by Q. Futurity, presentness and pastness, however, are not properties E possesses per se; these are relationships E has to M_1, M_2 and M_3. It is not contradictory to assert that E is related differently to different moments in time. This reply is clarified by pointing out that (a), (b) and (c) are equivalent to:

E is after M_1 . . . . (a')

E is simultaneous with M_2 . . . . (b')

E is M_3 . . . . (c')

It is clear that (a') (b') and (c') are not incompatible, for (b'), for instance, is not denying anything that has been affirmed by (a'). Thus, with the introduction of (a') (b') and (c') we have removed the difficulty we thought existed with (a) (b) and (c).

Once the argument is presented in this manner, Broad's objection disappears. In presenting the contradiction it was not overlooked that 'E is in the future' and 'E is in the present', etc., were not asserted at the same time. But although the first is asserted at M_1, the second at M_2, while the third statement is asserted at M_3, they seem to be attributing different properties to the same instantaneous event. Hence it would seem that only one of them could be true. We have agreed, after all, that (d) and (e) cannot together be true if P stands for a particular which has no temporal extension. Such a particular cannot undergo any changes, because whereas change consists in a particular having one property at one point in time and a different property at another point of time, P occupies a single point in time only. But so long as 'E is in the future' is understood to be asserting that E has a certain property, and 'E is in the present' as denying that this same property is possessed by E, while 'E is in the past' is understood to assign yet a different property to E not assigned to it by the previous two statements, then these three statements must be mutually incompatible even though they are made at three different moments). Consequently, (a) (b) and (c) must be mutually incompatible. Thus, definitely there is something 'in the temporal appearances to suggest that there is a contradiction to be avoided.' The contradiction is avoided by showing (a) (b) and (c) to be equivalent to (a′) (b′) (c′). Hence, we are forced to take the first step along the regress.

What about the next step? Now (a′) (b′) (c′) are B-statements that present no difficulties because they refer to permanent relationships between events and dates. (a′) and (b′), for example, do not contradict one another, because (a′) is not asserting something that is denied by (b′). (a′) asserts that a given relationship exists between E and M_1, while (b′) asserts that another relationship exists between E and M_2. It is A-statements that give rise to difficulties, because some of them assign properties to instantaneous events or dates which are denied by others. Are we to conclude perhaps that there are no temporal statements at all except B-statements? McTaggart denies such a possibility, for if there were only B-statements there could be no real change; and change is essential for the reality of time. We have eliminated the A-statements involving E by showing (a) (b) and (c) (which themselves are not A-statements but which mention A-statements) to be equivalent to (a′) (b′) and (c′), but we can reintroduce these if we wish by referring to the A-determinations of M_1. We have the statements 'M_1 is in the future', 'M_1 is in the present', etc., which may be all true if asserted at M_4, M_5, and M_6, respectively. These statements exhibit the changes M_1 may undergo. At the same time, however, these statements also assign different properties to M_1, which occupies a single point in time only, and hence cannot be seen to be able to accommodate more than one property. So once more the difficulty of incompatible statements arises, and we are forced to take a further step along the regress to resolve it.

We should now be able to see clearly how OBJ. 2 raised by Prior disappears. He is right of course that once we have eliminated the difficulty raised in stage one and have moved to stage two, we can stay there facing no contradiction. Yet if we did so we would be facing a new difficulty, namely, lack of change. But as we have seen in the previous chapter, A-statements are vitally important for McTaggart, for in their absence there is no possibility for change, and time would not be real if it did not permit change. This forces us to move to stage three, where once more we introduce change but where at the same time we land in a contradiction. Thus, although it is true that between every two stages along the regress in which a contradiction is involved, there is an intermediate stage free of contradiction; nonetheless, if we stopped at any of these stages we would be left with no facts that constituted genuine change. The regress we are caught up in is vicious in a remarkable sense: the difficulty we encounter varies from an odd to an even step. At each step along the regress, as soon as we escape contradiction, we are left without transient facts; and as soon as we take a step to restore change in the universe we meet with a contradiction.

Nor should it any longer be difficult to see why OBJ. 3 is also not valid. 'O is here' asserted at L_1, and 'O is there' asserted at L_2 (simultaneously), obviously do not contradict one another, for the second statement does not deny what is affirmed by the first statement. 'O is here' asserted at L_1 is equivalent to 'O is at the same place where θ is asserted', where 'θ' is the name of the statement 'O is here' asserted at L_1. The truth-value of this latter statement obviously does not vary with spatial position, for if it is true that θ and θ's assertion are co-spatial, then one is making a true assertion no matter where one asserts this to be a fact. On the other hand 'O is there' asserted at L_2 is equivalent to 'O is not at the same place where τ is asserted', where τ is the name of 'O is there' asserted at L_2. Once again, although 'O is there' may seem to vary in truth value from place to place, it is obvious that its logical equivalent—asserting that O and L_2 are not co-spatial—is true when asserted anywhere if true when asserted somewhere. Once this fact is pointed out we need go no further, for no additional problems exist. We need not be disturbed in the least by the fact that there are no spatial statements whose truth values vary from spatial position to spatial position parallel to temporal statements whose truth values do change from temporal position to temporal position. In other words, among spatial statements we do not have anything to parallel temporal A-statements.

McTaggart is quite happy with this circumstance for he never claimed that change is the essence of space. However, because change is essential to time, he does not permit us to analyse 'E is now' asserted at t_1 as being equivalent to 'E is simultaneous with θ', where 'θ' is the name of 'E is now' asserted at t_1. The first statement assigns the property of presentness to E, a property not permanently possessed by it. Consequently, 'E is now' changes truth-value depending on at what point in time it is being asserted, which converts it to an A-statement. It cannot be translated into a B-statement, such as 'E is simultaneous with θ', which if true is always true and if false is always false, irrespective of the time of its assertion. Thus, it is only with respect to time that an inescapable dilemma exists, in which we admit either an inconsistency arising out of the fact that we assign incompatible properties to a non-extended event, or we eliminate the inconsistency at the cost of eliminating change as well. In the case of space, we are quite content to do the latter and we do not mind that there are no spatial statements whose truth-values vary with the point in space where they are asserted.

8   The Three Types of Regresses

The most remarkable feature of McTaggart's effort to show that A-statements give rise to a fatal problem is that his underlying argument relies on an extraordinary regress. In general two types of infinite regresses have been recognized:

(1) *Vicious Regress.*   Suppose the following has been asserted:

A theorem is not understood unless it is derived logically from a theorem that is understood.

Let us say that we are presented with theorem T_1 and succeed in showing that it follows logically from theorem T_2. Can we say that we now comprehend T_1? Not yet, of course, because it has been shown only that it follows from T_2, and T_2 itself is not understood until it is derived from an intelligible theorem. Suppose we show that T_2 itself follows logically from T_3; in fact we succeed in showing that.

$$T_n \rightarrow T_{n-1} \rightarrow \ldots \ldots \rightarrow T_4 \rightarrow T_3 \rightarrow T_2 \rightarrow T_1$$

Unfortunately, all our efforts will have been in vain. That is to say that in spite of all the derivations, T_1 remains just as unintelligible as it was in the beginning. This is so because T_1 has not been shown to follow from an understood theorem; it has been shown only to follow from T_2, which is incomprehensible because rather than derived from an intelligible theorem it has been derived from T_3, which is unintelligible because, . . . etc., up to T_n, which has not been shown to follow logically from any theorem and thus is unintelligible. Our original problem was the comprehensibility of T_1, and we have contributed nothing to its solution through our demonstration that there is a chain of theorems beginning with T_n which lead to it. Thus, no matter where we stop along this regress, which may go on to infinity, matters remain as they were before we had taken even our first step. This is therefore a vicious regress.

(2) *Benign Regress.* Consider the dictum 'Every event has a cause'. It follows from it that event E_1 must have a cause. In answer to the question what is the cause of E_1, we may discover that it is E_2. Later we may find that E_3 is the cause of E_2 which itself is caused by E_4, and so on. According to the dictum, this chain of causes continues ad infinitum. Suppose we have proceeded as far as E_n, which we had found to be the cause of E_{n-1}, which is the cause of . . . . E_4, and we stop there. In this case, of course, we would not know the cause of E_n but we would know the causes of E_{n-1}, ... E_4, E_3, E_2 and E_1. Proceeding along the regress did make a difference; we know more now than before. Not only have we solved our first problem, namely, what is the cause of E_1, but we know also the answers to the questions what are the cause of E_2, E_3, E_4, ... E_{n-1}. This is therefore a benign regress.

The main difference between (1) and (2) is that in case (1), no matter where we stop along the regress, we face the same difficulty which made us start along the regress in the first place. In the case of (2), however, each step we take along the succession of particulars that form our infinite sequence is beneficial. The further along the regress we stop the better off we are. (v. Q&A 40)

Through McTaggart, however, we have now discovered a third type of regress:

(3) *Alternating Regress.* In McTaggart's case, as in (2), it is useful to take the first step, for by doing so we solve the problem that has originally troubled us, namely, that incompatible properties cannot attach to a temporally unextended particular. However this step lands us in a new difficulty, in which our A-statement turns out to be a B-statement, and thus we have lost the statement that expresses change. By taking the next step we solve this difficulty but are saddled once more with the original difficulty, and so on. Each step along the regress solves one of our difficulties but gives

rise to the other. Thus, no matter where we stop along the regress, we are faced either with our first or our second problem. This regress is therefore fatal too.

One of the ironic misunderstandings on the part of Russellians who attack McTaggart's argument as being fallacious—a misunderstanding that has persisted throughout most of this century—is caused by their failure to see that what McTaggart has done is what they themselves have been endeavoring to do. McTaggart, like Russellians, shows that A-statements are basically incoherent. He argues that if the statement that assigns futurity to E and the statement that assigns presentness to it are both timelessly true, then we are faced with a clear contradiction. What he will not settle for is their solution, namely, that these statements be turned into B-statements, and we resign ourselves to the fact that the temporal series consist of moments and events which have no more than B-properties. (v. Q&A 41)

IV

BACKWARD CAUSATION AND THE QUESTION OF THE STATUS OF THE FUTURE

1 *The Humean View of Causality*

An inquiry into the question whether an effect may precede its cause, that is, whether backward causation is possible, may yield important results. Among the things to which it may contribute is a greater understanding of the nature of time. Let me just indicate how. According to some philosophers a fundamental difference exists between the status of the past and the future. Admittedly, events of the past do not exist now but only in the past. Nevertheless, the statement that an event has happened in the past is true in the present too, if it is indeed a fact that the event in question did take place. However, the statement that a given event will occur, even if it should turn out later that the event did occur, has no truth-value in the present. This claim is based on the argument that unless we maintain that statements concerning future contingencies are of indeterminate truth-value, we are forced to subscribe to fatalism. But most people when faced with the choice between accepting fatalism, according to which we are entirely powerless to affect our future, and renouncing the truth of the traditional law of excluded middle, according to which every proposition is either true, or if not, then false, will naturally choose the latter alternative.

The majority of philosophers, however, have rejected this contention. They have argued that propositions about future contingencies may well have truth-values in the present, yet fatalism does not follow. I shall leave aside the extreme view which holds that such a fundamental difference exists between the past and the future and look now at a more moderate view which maintains that a weaker difference exists. This view, although maintaining that statements concerning future events have a determinate truth-value, does not consider that it necessarily follows that the future is real now in the same sense that the past is real. The way in which past and future stand to the present may yet be dissimilar, albeit in

59

a weaker sense. It may be claimed that although past events exist in the past alone, their influence is often felt in the present; whereas future events are incapable of influencing the present. The past certainly has not vanished without a trace; in a sense it is still with us, shaping the form of present occurrences. Ongoing events are dominated by past events and thus lend continued reality to the latter. The reality of moments receded from our presence manifests itself now in the role those moments play in determining the character of present moments. On the other hand, the present is completely free from being dominated by any event yet to come. The future has no reality by means of which to reach out toward us and make an impact on the present.

This view would be dealt a serious blow, however, should we discover that cases of backward causation do in fact take place. It would be very difficult to maintain the unreality of the future when events not having happened yet have the power of bringing about certain occurrences in the present.

But the question is, is it at all possible to make such a discovery? According to some philosophers, it is not. If their position is correct, then the interesting view that holds that the future is less real than the past in the sense just mentioned loses substance because it has become irrefutable in principle. The proposition that causes always precede their effects is trivially true and thus has no relevance to the question of the unreality of the future to any degree.

The discovery of a case of backward causation is indeed trivially ruled out by the Humean view of the nature of causality. Hume says in his *Treatise*:[1]

> The second relation I shall observe as essential to causes and effects is not so universally acknowledged but is liable to some controversy. It is that of *priority* of time in the cause before the effect.

According to Hume when event-types are constantly conjoined a causal relation exists between them, and the event that precedes the other is the cause. On this view backward causation is ruled out by definition.

Thus, backward causation is possible in principle only if causes are assumed to be intrinsically different from their effects. It is required that independent of its temporal relationship to its effect, the event designated as the cause has some distinct characteristics,

1. David Hume, *Treatise on Human Nature,* ed. by L. A. Selby-Bigge, Oxford, 1975, p. 76.

by virtue of which we recognize it as the event which brings about the occurrence of the other event, produced by it as its effect.

A number of serious attempts have been made to characterize causes and effects in terms other than temporal priority. Notable among these is the 'manipulability theory of causation' advanced by Gasking and Von Wright. Another attempt is that of J. L. Mackie:

> If events x and y are causally connected in a direct line then x is causally prior to y if there is a time at which x is fixed while y is not fixed otherwise than by its causal connection with x.[2]

Mackie endeavors to explain the notion of 'fixity'.

These attempts have faced various criticisms, including that of Alexander Rosenberg in a recent article "Propter Hoc, Ergo Post Hoc",[3] in which he concludes that all efforts to define causes and effects, independent of temporal priority, have failed, and the correct position is that of Hume. Thus, we would have to conclude that the many studies in the past concerning whether effects may precede their causes are devoid of substance.

In what follows I shall not be trying to restore any of the arguments of the past designed to distinguish between causes and their effects. Rather, I will consider three new suggestions as to how to determine which of two events is the cause and which is the effect. The first two are open to attack. The third, however, which differs sharply from any of the known suggestions made in the past, I believe to be a tenable suggestion.

2   Causes as Explanations

I shall begin by discussing an attempt to distinguish between a cause and its effect based on the asymmetry between them, which is manifested by the fact that by citing a cause we can explain why an effect has taken place but not vice versa. I also exploit the fact that not only a physical event may be explained by citing the causes that brought it about, but a human act, too, can be explained if we can describe the reasons taken into account by the agent that might have induced him to act in the way in which he did. Now a situation may arise in which a human act A is causally connected to a physical event B. But if we look upon B as the cause, then we can explain why A occurs without being able to explain why B occurs.

---

2. "Causes and Conditions," *American Philosophical Quarterly,* 1965, p. 257.

3. *Am. Phil. Quart.* 1975.

On the other hand, if we take A to be the cause, then we can explain why B occurs, namely, because of A; and A is not left unexplained because there happen to be good reasons why a rational agent should do A. In this case we should look upon A as the cause, so that we have a fuller explanation of the phenomena before us. If we do so then in cases in which A occurs after B, a cause will be said to occur after its effect.

An illustration is provided by the following story. A man in the Himalayas, who is considered to be a wizard by his countrymen, hears that some disaster, such as an earthquake, hurricane, flood, fire, or the like, has occurred anywhere on earth, and, at will, he names a sum that always turns out to equal the damage caused by the disaster. One might be tempted to say at once that here is a case of backward causation, as the amount of damage is confirmed to be dependent on the amount named by the wizard, and the amount named by him is whatever he freely decides he wants to name. In reality, however, no grounds exist yet for saying that we are confronted here with a genuine case of backward causation. Although the amount of damage seems to be manipulated at will by the wizard there is no reason why one should not claim that it is the occurrence of an event with a certain magnitude of damage that determines that the man in the Himalayas will want to name an amount of that magnitude. Although he may feel no constraints at all and says whatever he wants to say, there is no obstacle to claiming that the amount he will want to name is entirely determined by the events which occurred before.

Suppose, however, that scientists become interested in this extraordinary phenomenon and decide to investigate it more closely. For this purpose a group of them organize an expedition to the Himalayas. They come equipped with their radios and listen to reports about disasters which occur at various places and note the amount of damage reported. In order to test the wizard more severely, they do not reveal to him the amount of damage disclosed over the radio but only that a disaster occurred at such and such a place. Then they ask him to name a sum different from the sum they have heard. The magician obliges and names the sum requested by the group of scientists. To everybody's great surprise there is soon an announcement that the original report was based on a mistaken estimate, and the actual damage is the exact amount named by the wizard.

Under these circumstances and if we still insist on giving an account of all that has happened in terms of forward causation, we could say that the amount named by the magician is determined by the amount he is requested by the scientists to name, and the amount named by the scientists is determined by the actual damage which has occurred. We would, however, have no explanation as to why there has been a misreporting of the actual damage and why such misreporting has become a regular occurrence since those scientists arrived on the Himalayas, whereas in previous times a mistaken reporting has been a most rare occurrence. On the other hand, if we are prepared to explain phenomena in terms of backward causation, then we can account more fully for what is going on. We can, of course, explain that the magician names the sum he names because he was requested to do so by the scientists, to whom he is anxious to demonstrate his special powers to determine the amount of damage done. We explain the fact that the scientists name a sum different from what they have heard on the radio on the basis of our knowledge that they want to create special conditions under which they can test more severely the skills of our man in the Himalayas. Thus, we have a reasonable explanation why in recent days the amount of damage caused by public calamities is misreported so regularly. This is simply the result of the fact that the scientists name, for their own legitimate and explained purposes, a sum different from that which was originally reported over the radio; and the magician, who ultimately determines the actual sum to which the damage amounted, agrees (in order to prove his efficacy) to name the sum requested by the visiting scientists. If we subscribe to the principle that when alternative hypotheses present themselves we have to select the one which explains more fully the data, then it follows that we have to maintain in the situation described that backward causation has taken place, for by doing so we manage to explain the situation more fully than by the alternative hypothesis.

It seems, however, that one could object to our conclusion by insisting that the principle called on to increase the range of our explanations does not apply in cases in which the hypothesis that explains more is a bizarre and very improbable one. It would defeat the purpose of the principle if it applied to such cases. After all, that purpose is to leave less facts unexplained, but by introducing an outlandish hypothesis we leave unexplained why such an unlikely hypothesis should be true. It may therefore seem that we could object and say that while the hypothesis, according to which we are facing here a case of backward causation, succeeds in explaining the otherwise unexplained anomaly of damages being systematically misreported, it leaves unexplained how such a strange and unlikely situation arises in which later events are the causes of effects preceding them.

Someone particularly keen on the idea of backward causation

might contend that such an objection is not well-founded. There is no basis for claiming—he might say—that backward causation is a bizarre and unlikely hypothesis. After all, it has not been demonstrated that such a hypothesis may not apply elsewhere. We have agreed that all we can say about the vast majority of cases is that they represent causally connected events; and unless we arbitrarily designate one as the cause and the others as the effect, we have no way of telling which is the cause and which is the effect. Thus, the majority of cases provide no positive grounds for claiming that an effect has preceded its cause, for they are just as well described as cases in which the cause has occurred before its effect. Hence, we have no definite proof that backward causation does not take place often, except for the fact that we lack proof that it does. In our example, where there are good reasons to maintain that such a state of affairs has obtained, we should be permitted to assume that backward causation is occurring, for there is no adverse evidence from elsewhere forbidding this conclusion.

However, one may raise another objection, as follows: granted that when we postulate forward causation in our story we do not know how to explain why damages lately are always misreported, nonetheless, we have no definite proof that there are no prior causes which happen to ensure that in all cases in which misreportings have occurred, misreportings had to occur. It could be maintained that whenever damages are wrongly assessed, there are characteristic peculiarities present in the physical situation which engender the mistake. If we only knew exactly what these peculiarities were we could well explain, in terms of prior causes, why there have been so many mistaken reports.

The philosopher sold on backward causation might wish to argue that this objection is not valid either. It is a general principle to choose the best available explanation and not to reject it by saying that there may yet be a better one, which we do not know yet but which is correct. If such an explanation should be forthcoming, then we should, of course, give up our presently held one, but until then the best explanation and the one we should hold to is that we have here a case of backward causation.

It would nevertheless seem that the strangeness of the situation described is such that if it were to occur one would be reluctant to consider it a case of backward causation.

3   *The Role of Interference*

Another attempt to distinguish between a cause and its effect is involved in making use of the asymmetry introduced by an interference. An interference with a cause interferes with its effect as well, whereas an interference with an effect need have no impact at all on the cause. For example, two kinds of events, A & B, may be symmetrically related to one another so that A is a sufficient and necessary condition for B, and vice versa, as long as there is no active interference directly preventing either of them from happening. However, when one of these events is interfered with an asymmetry arises, so that an interference with A, which stops it from occurring, always results in B not taking place either; but an interference that directly stops B from occurring does not necessarily prevent A from occurring. This feature in their relationship dictates thinking of A as the cause and B as the effect. For even though normally B is sufficient and necessary for A, it is not the case that B causes A, for A is the kind of event that may occur spontaneously and uncaused (as evidenced by the fact that when we directly interfere with B, A may nevertheless sometimes occur). The following table illustrates our case by showing what combinations of events are possible and what combinations are impossible:

	A occurs	Interference	B occurs	
1	Yes	None	No	Impossible
2	No	None	Yes	Impossible
3	No	with A	Yes	Impossible
4	Yes	with B	No	Possible

Case 1 is impossible, because A is sufficient for B and B is necessary for A. Case 2 is impossible because B is sufficient for A and A is necessary for B. Case 3 is impossible for the same reason. Case 4, however, is possible, because A is the kind of event that can occur uncaused. Normally, its occurrence is sufficient for the occurrence of B, as shown by Case 1; but when B is prevented from happening then A, which is independent of B, can still occur, but B, of course, cannot.

An illustration may be served by two pendula, *a* and *b,* so isolated from their surroundings that nothing will set them off to swing. Sometimes *a* and *b* both begin to move spontaneously but never one without the other. We shall denote *a*'s beginning movement as event A, and b's beginning movement as event B. Whenever A happens, B happens, and vice versa. Suppose the bob of a is fastened so that it cannot move; consequently, we never observe b moving although it is free to move. On the other hand, when we render b immobile, this action interferes only with b's movement but not with the movements of a. In such a situation, it would seem as if we could assume that b's oscillation was caused by a's oscillation, rather than the other way around. Furthermore, if we suppose that B always occurs a few moments before the occurrence of A, then we have a full description of a situation in which an effect precedes its cause. The fact that no logically possible situation of this kind is known ever to take place may be said to support to a certain degree belief in the physical impossibility of backward causation.

It seems, however, that if someone wanted to resist the conclusion that backward causation occurred in the situation described above, he could do so. He would redescribe several aspects of the situation and call B the cause and A the effect. Why then is Case 3 impossible? Because what we have called an interference with A is in fact also an interference with B. It is not, after all, entirely unreasonable to claim under these conditions that we regard the anchoring of a pendulum a as interfering with the movement of b. If such anchoring also stopped b from swinging, we might be inclined to look upon the tying down of a as an interference with the motion of b. To the question why Case 4 is possible, we might answer that, although B is the cause of A, it is only a sufficient but not a necessary condition for A. What is necessary for A is either B or the occurrence of the event which we have described as 'the interference with B'.

In the next three sections I shall put forward what I believe may be an unobjectionable suggestion as to how we can observe which of two causally connected events is the cause and which is the effect. I shall not proceed by using any of the arguments of the past designed to distinguish between causes and their effects. My own attempt to show how we may determine which of two events is the cause and which is the effect will differ sharply from previous attempts, in that I shall not proceed by offering a definition of a cause. Given two events E_1 and E_2, there may be nothing in the characteristics of either, as such, which will reveal which is the cause. Thus, looking at the two events alone, permits us to say with equal assurance that E_1 causes E_2, or that E_2 causes E_1. However, when we study the surrounding circumstances, they may be seen to provide evidence that E_2, which happened later than E_1, is the cause of E_1. The circumstantial evidence may keep accumulating indefinitely until everyone is forced, under its weight, to admit that E_2 must indeed be the cause of E_1.

Contrary to the opinion of some philosophers who maintain that by closely examining the features of an event one may observe it to be the cause of another event to which it is linked, I shall maintain that the assertion that E_2 is the cause of E_1 is not to be made as a result of carefully watching E_2 or E_1. It is an hypothesis for which one may obtain any amount of evidence from a variety of sources. The amount of evidence eventually may become so overwhelming that even a person who has the strongest resistance to the idea that an effect might precede its cause may not be able to withstand acceptance of the hypothesis. In what follows I shall compare the hypothesis that E_2 causes E_1, and not the other way around, to the hypothesis that the earth is round and not flat. The truth of the latter hypothesis cannot be directly observed either, yet even those who in the past were most strongly wedded to the idea that the earth was flat, had to yield to the prodigious amount of evidence supporting the round-earth hypothesis.

4   The First in a Series of Thought Experiments

I propose now to perform a number of thought experiments. In these, various laws unheard of today will be assumed to hold. These experiments will have to obey the sole restriction imposed on all such experiments: their description must not contain a contradiction.

Let a = X is magnetized at p

b = Y emits radiation at p

α = X rotates at p

β = There is an electric field at p

We shall suppose that on a large number of days it has been observed that whenever object Y which is located at p emits radiation at a certain time of day t_1, then at a later time t_2 there is an electric field at the same place. This may be briefly expressed as

(i) b at t_1 → β at t_2

We shall also suppose that it has been established through repeated observations that in case X rotates at t_2 at p in an electric field then

X is always magnetized at t_1. This can be expressed as:

(ii) α at t_2 & β at $t_2 \rightarrow a$ at t_1 ($t_2 - t_1 = 100$ seconds)

We shall now consider at least two interpretations of what is taking place. Interpretation B, which will turn out to be the more straightforward interpretation, permits us to assume that backward as well as forward causation is taking place. On the other hand, interpretation F, at the cost of being more encumbered, assumes that only forward causation is taking place:

Interpretation B: We explain (i) by saying that b represents a sufficient cause of β. Thus, (i) does not imply any backward causation. However we explain (ii) by claiming that whenever both a and β obtain, that is, whenever it is true that X rotates at t_2 at p and there is at the same time an electric field at the same place, then as a consequence X is caused to be magnetized at t_1. That is, X's rotation at t_2 in an electric field, results in its becoming magnetized at an earlier time t_1.

Interpretation F: We explain (i) of course in the same manner as before. To avoid the assignment of backward causation to any phenomenon we shall explain (ii) in the following manner: from α & $\beta \rightarrow a$ it follows that $\sim a \rightarrow \sim (\alpha$ & $\beta)$ which may be taken to say that a is a necessary cause for α and β to be jointly true. We have to say that a itself is caused by some event E_1 occurring prior to t_1 and which we cannot identify. According to Interpretation B we are not faced with a similar difficulty because the cause of a is given by (ii) and the cause of α may be assumed to be Tom who at t_2 gets hold of X and starts rotating it.

Now to an important point. We know that α & $\beta \rightarrow a$ implies $\beta \rightarrow (\alpha \rightarrow a)$ which is perfectly acceptable according to Interpretation B, for it means that when β is true then α backwardly causes a. The followers of Interpretation F however would have to say that this is true basically because $\beta \rightarrow (\sim a \rightarrow \sim \alpha)$, that is, not because α is a sufficient cause for a but rather because a is a necessary cause for α. What we may say then is that when β at t_2 is true, then a at t_1 is required for α at t_2 to obtain. In other words if Y emits radiation at t_1 at p then X will not rotate at t_2 at p unless X is magnetized at t_1. This restores forward causation.

But there is a problem here, to wit: how can a be necessary for α when it has been clearly observed that it is Tom, who holds X in his hands, who represents the source of the force bringing about the rotation of X? Surely Tom's action is sufficient by itself to make X rotate—if we assume Tom to be a strong man and X a relatively small object.

In order to save hypothesis F it will have to be postulated that perhaps Tom, who does not in the least feel that he is being compelled against his will to do what he does and who may not at all be aware of the fact that X is magnetized at t_1 is actually caused by a to decide to rotate X at t_2. In other words $b \rightarrow (\sim a \rightarrow \sim \alpha)$ is taken to show that in case Y emits radiation at t_1 at p, then unless X is magnetized at t_1, X will not rotate at t_2 at p. This in itself puts a considerable strain on the present interpretation for it would be very surprising, and contrary to what we would expect from our background knowledge of psychology, to find that brain events should have necessary causes like the one represented by a . Further difficulty arises from the fact that Tom seems to be able to rotate at will every object in sight, with the apparent exception of X, without any of those objects having been magnetized 100 seconds prior to his action. Thus in general the rule seems to be that Tom's decision to rotate an object does not require the magnetization of that object 100 seconds earlier.

Thus faced with the facts expressed by (i) and (ii), we may postulate either hypothesis B or hypothesis F. Hypothesis B fits considerably more smoothly in our system of knowledge: a is caused by α and β; β is caused by b, and α is caused by Tom where Tom's own act is either freely willed in the sense that it has no causes at all or is caused by antecedent brain events. On the other hand according to hypothesis F we simply may not be able to identify the cause of a. It is to be assumed that some physical event E_1 occured prior to t_1 which brings about a at t_1. Furthermore as we have just mentioned it is also necessary to assume some very strange things concerning the genesis of the mental event constituting Tom's decision to rotate X at t_2.

Suppose that we decide to perform an experiment to further test the relative merits of hypotheses B and F. Let it be given that prior to January 1979, on some days a obtained at t_1 and on some days it did not. On those days on which a did not obtain at t_1 neither did α obtain at t_2, provided β obtained at t_2, as was to be expected from $\beta \rightarrow (\sim a \rightarrow \sim \alpha)$. We have seen already how this is accounted for differently in Interpretation F and in Interpretation B. Now let us suppose that on the last day of 1978 we promise Tom that if he will start rotating X at t_2 and keep rotating it for 10 minutes, he will receive for his services $100 every day for the next month. Given Tom's poor financial situation and the easiness of the task he is being asked to perform, it would seem almost a certainty that Tom would be very eager to comply; and hence we would expect that during the month of January a will obtain at t_2 every day.

It is not entirely clear, however, that a scientist subscribing to Interpretation F would make such a prediction. According to him a at t_1 is a necessary condition for a at t_2. And as on some days a does obtain at t_1, whereas on others it does not, it should follow that only on some days will Tom rotate X at t_2. On the other hand a scientist may say that whereas we have evidence that under ordinary circumstances a at t_1 is required for a at t_2, in our case, where strong incentive has been provided for Tom to rotate X at t_2, we see that he will decide to do so even in the absence of a at t_1. Or, in other words, (ii) will no longer hold. Be that as it may concerning the question whether a is at t_2, the aforesaid scientist will certainly have no reason to expect that a be at t_1 on every day, for in the past a has failed to occur at t_1 on as many days as on which it did occur; it is assumed to be depending on whether E_1 occurs prior to t_1, and we may infer from past experience that on some days it does and on some days it does not. We have no grounds on which to assume that something had occurred to interfere with E_1.

Suppose now that January has passed, and we have observed that a at t_1 and a at t_2 on every day. This presents us with nothing unusual—nothing that would require any special postulates if we subscribe to Interpretation B. Why is a at t_1 on every day during January 1979? Because a at t_2, which in itself constitutes sufficient reason for a at t_1. Then again the fact that a at t_2 on every day this month is accounted for simply by the promise made to Tom. Tom, becoming aware of our offer on December 31, 1978, which is a mental event, ensures that the mental event of his deciding to rotate X at t_2 on every day during January 1979 will take place, and his decision is sufficient reason for the physical event of X actually to be rotated to take place.

On Interpretation F, however, a is not caused by the later event a but rather, as we have said, by an earlier event E_1. Prior to 1979 on some days a ocurred at t_1, and on some days it did not. We are to assume that on some days E_1 did take place, and that on others it did not. During January 1979, however, we are compelled to say that E_1 takes place on every day. We are thus faced with the difficulty of how to account for the fact that on all those days for which Tom had been promised \$100 for rotating X at t_2, the event that brings about the magnetization of X at t_1 invariably happens to take place. In order to overcome this difficulty some ad hoc hypothesis is required. An adherent of Interpretation F could straighten out matters if he postulated that the mental state of Tom, prior to t_1 each day in which he is aware that if he rotates X at t_2 he will receive \$100, is a sufficient condition for E_1 to occur. Thus we see that one cannot maintain Interpretation F unless one is prepared to subscribe to a number of strange ad hoc hypotheses.

I do not wish to insist that this evidence is sufficient to elicit everyone's admission that we are faced here with a case of backward causation. This is not how unrelished hypotheses are embraced. The reluctance to entertain the thought that effects may precede their causes may be too great and prevent at this stage admission by most people that cases of backward causation exist. To use an analogy: in early antiquity most people considered it far more reasonable to assume that the earth was flat rather than round, a judgment which seemed to be supported by common sense and common observation. The absurdity of the hypothesis that the earth was round seemed evident from a number of considerations, e. g., if the earth were round people at the antipodes would be standing upside down. But then people because aware of the relevance of the fact that the bottoms of receding ships disappeared over the horizon before their upper parts. This observation did not convince everyone to relinquish the flat-earth hypothesis although it made the defence of it considerably more difficult. For as history demonstrates, cherished hypotheses are abandoned reluctantly. Only an ever-increasing and ultimately overwhelming accumulation of evidence — which clashed with the flat-earth hypothesis, in consequence of which that hypotheses could only be kept afloat with the continual addition of extra hypotheses specially designed to prevent it from sinking beneath the increasing weight of prima facie hostile observation, but which themselves piled up a great amount of ballast on the favored hypothesis thus causing it to sink — caused men of reason to accept as sufficiently proven the theory that the earth is round.

In a similar manner, rather than abandoning their view, adherents of Interpretation F may come up with an ad hoc hypothesis postulating that Tom's awareness on any given day that he can earn \$100 by rotating X at t_2 has a causal connection to a which it brings about. On Interpretation F one has to postulate an unusual kind of causal connection between Tom's awareness and a at t_1 as well as between the latter and Tom's decision to rotate X at t_2. Even if this were not strange, Interpretation F is at a disadvantage of having to postulate two extra causal connections which were not predictable. We shall describe how further difficulties may pile upon each other so as to eventually force those who cling to Interpretation F to give up their cherished hypothesis. (v. Q&A 44, 45, 46, & 47)

5 The Second Thought Experiment

Let us suppose that Z is an object that when placed in an electric field counteracts and neutralizes the field. Our second experiment consists in watching what happens when Dick places one day at t_2, Z at p. Suppose we observe now that $\sim B$ at t_2. This observation creates no difficulty in spite of the fact that b obtained at t_1, and we have been given that (i) b at $t_1 \rightarrow \beta$ at t_2. What we shall establish is that (i) is true under normal circumstances, that is, when there is no positive interference preventing β to take place at t_2. By placing Z at p, Dick causes the electric field generated by Y to be cancelled through the influence of Z.

With respect to the question whether a obtains at t_1, the expectations of those who hold Interpretation F will differ from those who hold Interpretation B. On the first interpretation, in case the conditions prior to t_1 seem to be the same today as in earlier days when Z was not introduced, there is no reason to assume that X is not magnetized at t_1. Those who hold Interpretation B, on the other hand, hold that α and β at t_2 are the events that bring about a at an earlier time t_1. As today β is not occurring at t_2, and in view of the fact that no new event has occurred which looks as if it may be an alternative cause for a at t_1, Interpretation B would expect X not to be magnetized at t_2. Of course a at t_2 is ensured by Tom wanting to rotate X from t_2 onwards.

Now suppose that in fact we observe that $\sim a$ at t_1. This presents a new difficulty for Interpretation F. Today and on all subsequent days in which Dick introduces Z at p at t_2, the conditions prevailing prior to t_1 seem identical to those that prevailed when Z was not brought into the picture. It is E_1 that brought about a at t_1, and nothing seems to have happened prior to t_1 which would warrant us to expect that on these days E_1 does not take place. Also, how are we to explain the seeming association between $\sim a$ at t_1 and $\sim \beta$ at t_2? The latter cannot be the cause of the former, because that would amount to backward causation; nor can the former be the cause of the latter, for we know that $\sim \beta$ at t_2 is caused by the introduction of Z at p.

In order to keep afloat Interpretation F, some new and quite strange ad hoc hypothesis is required. One such hypothesis would be that event E_2, which we designate as the antecedent cause of Dick's decision to introduce Z, also happens to counteract E_1, the event that would otherwise cause X to be magnetized at t_1. Of course, we may never succeed in identifying E_2; yet we postulate it to have occurred some time before t_1 and assign to it causal powers in two very different areas: E_2 causes the mental event of Dick deciding to introduce Z and also it causes the neutralization of E_1 and thus prevents X from becoming magnetized.

By now hypothesis F begins very much to look like the flat-earth hypothesis. As is known, by the time of Pythagoras people came to the realization that the fact that lunar eclipses are round, together with the fact that they always occur at the middle of the lunar month when the line sun–earth–moon may be straight line, and the undisputed fact that the shape of shadows resembles the shape of the objects throwing them, strongly indicated that the earth is round. The flat-earthists could hardly deny that lunar eclipses looked circular. Thus, they were forced to invent some alternative ad hoc theory of lunar eclipses, according to which these occurrences were not caused by the obstruction of the sun by the earth. This maneuver, of course, left their cherished hypothesis heavily encumbered, for among other things they were hard pressed to explain why lunar eclipses always occur exactly in the middle of the lunar month. Nevertheless, determined flat-earthists were ready to tolerate such difficulties and hold on to their pet theory, rather than to accept what appeared to be an unpalatable alternative. But they could not do so indefinitely, for in the course of history their theory continued to be subverted. At some stage travelers proceeding along a straight line found themselves eventually returning to their point of origin. This supported very strongly the claim that the earth may be circumnavigated because it is round. Very complex and unlikely extra hypotheses were required on the part of flat-earthists to dismiss this phenomenon and to describe the situation in such a manner that in fact no circumnavigation of the globe had taken place. Thus, the flat-earth hypothesis became more and more unwieldy, until even its most devoted supporters could not keep it from going under. With the advent of space travel, photos of the earth taken from outer space show it to have a spherical shape. Perhaps even in the face of this evidence, one could cling to the flat-earth theory, but surely the cost would be prohibitive. It is safe to say that by now all reasonable people are compelled to concede that the flat-earth theory must be abandoned. It is logically possible to construct a situation in which hypothesis F meets a similar fate.

6   The Third Thought Experiment

Let us imagine that at $t_2 + 10$ Harry destroys Z and in consequence we observe that β at $t_2 + 10$. This observation is accommodated equally well on both interpretations. From (i) we know that b at $t_1 + 10 \rightarrow \beta$ at $t_2 + 10$ and b at $t_1 + 10$ obtains, as Y emits radiation from t_1 onward at all times including $t_1 + 10$, and nothing is known to exist that would prevent the prevailing of an electric field at p at $t_2 + 10$, because Z has been destroyed.

Now we may ask ourselves, is X magnetized at $t_1 + 10$? It is evident that those who hold Interpretation F will say no. The reason is that nothing is known to have taken place prior to t_1 to bring about the magnetization of X at t_1. We know that E_1, which would be a cause for a at t_1, had taken place but this was neutralized by E_2, as we have said in the previous section. It has not been given that some new cause, immune to E_2, has occurred prior to $t_1 + 10$, and thus X should be assumed to continue in its unmagnetized state as before. Concerning the question whether X rotates at $t_2 + 10$, once more the answer is not certain. It may be no, because $\sim a$ at $t_2 + 10$, and a 100 seconds earlier is a necessary condition for α to occur provided β occurs. But it may be yes, for it is possible that this is so under normal conditions but in our case the promise of $100 is sufficient to ensure that α obtains for 10 minutes beginning at t_2. But irrespective of what the answer is to this question, X must be assumed unmagnetized at $t_1 + 10$, just as it is a second before that, as nothing has happened to bring about the magnetization of X.

Those who subscribe to Interpretation B, however, will say that X is magnetized at $t_1 + 10$. The reason according to them why X was not magnetized before is that there was no electric field at p at the appropriate time, and hence there were no sufficient causes for X to be magnetized. But because at $t_2 + 10$, Z is destroyed, in consequence of which β at $t_2 + 10$, it follows that at $t_2 + 10$ both α and β obtain (the former obtains because Tom rotates X from t_2 for 10 minutes). These are sufficient conditions to ensure that a at t_1.

Suppose that we observe that a at t_1, which is contrary to what one should expect on Interpretation F. Further ad hoc hypothesis would be required to save that interpretation. It might be suggested that there is some unobserved event E_3 occurring prior to $t_1 + 10$ which causes a at t_1 and also causes Harry to destroy Z at $t_2 + 10$. This, of course, would further complicate Interpretation F, which then could no longer be maintained unless three unobserved events were postulated as well as a number of ad hoc causal conditions were assumed to hold. The interpretations would differ greatly from one another in that one fits smoothly with everything observed whereas the other has been greatly encumbered by ad hoc hypothesis.

By now I believe the reader might be sufficiently convinced that we could go on indefinitely, inventing more and more experiments whose results fit very smoothly with hypothesis B. In fact, followers of hypothesis B would always readily predict the results of these experiments. At the same time each experiment would create added difficulties for hypothesis F, which could be saved only at the expense of burdening it with further and further ad hoc hypothesis specially designed for keeping it from going under.

Sooner or later we are bound to reach a point where the ballast accumulated by hypothesis F is so heavy that no reasonable person would henceforth make efforts to save it from sinking beneath the burden of its encumberances.

Thus, it is logically possible to construct a situation in which evidence is so overwhelmingly in favor of the hypothesis that backward causation is taking place that no reasonable person would withhold his concurrence. It is therefore not possible to safeguard by a simple definition the theory that causes should invariably precede their effects. It seems logically possible that a series of experiments could irrefutably indicate that as a matter of empirical fact we are faced with an instance of backward causation.

It also follows that a discussion of the philosophical implications of a situation in which backward causation would have to be said to have occurred, or what conclusions we could derive from a proof that such an occurrence shall never take place, is not devoid of significance. Contrary to the dictum of Hume and his followers, the possibility of backward causation cannot be ruled out simply by fiat.

7 Newcomb's Game

I.

Now that we have satisfied ourselves that in principle it is possible to discern cases of backward causation we naturally should like to go a step further and inquire whether in practice effects may precede their causes or whether for some reason this may never happen. In what follows I shall consider an attempt to prove that backward causation can never occur. That is, I shall attempt to show that even though effects can be distinguished from their causes and that if they do occur first we would be able to discern this, in principle it is impossible that an effect should precede its cause.

In order to do this I shall now discuss what has become known as Newcomb's Game. As I have dealt with this at great length in my *Religion and Scientific Method* (Dordrecht 1977) I shall be relatively brief and concentrate on some aspects of this topic not discussed in that book. I ask the reader to take special care in reading this section, because in the past even competent philosophers have shown strong propensity to get things wrong from the very start.

I shall begin with what I have called Game I. In this game a chooser (CH) is confronted at t_1 with two boxes, one of which is transparent and the other opaque, and he is allowed either of two choices:

C_1:   Choose the opaque box only.

C_2:   Choose both boxes.

The transparent box is seen to contain $1000. It is also given that 24 hours earlier at t_0, a person with a vast record of correct predictions, who therefore is presumed to be a perfect predictor (PR), seals the two boxes and

(A)   If PR predicts that CH does C_1 at t_1, he puts $1,000,000 ($M) in the opaque box.

(B)   If PR predicts that CH does C_2 at t_1, he puts nothing in the opaque box.

In *Religion and Scientific Method,* I explained that the perfect past record of the PR lends itself to basically two different interpretations. Interpretation 1, the widely known interpretation, takes the evidence as supporting the conclusion that by doing C_1, CH ensures P_1 (= PR predicts C_1 and places $M in the opaque box) and by doing C_2, CH ensures P_2 (= PR predicts C_2 and places nothing in the opaque box). On Interpretation 1 we are faced with a case of backward causation, and to the question what is better for CH to do, the answer seems to be C_1. The reason is because C_1 and P_1 results in obtaining $M contained in the opaque box, whereas C_2 and P_2 results in getting $1000 only. Of course C_2 and P_1 would result in obtaining both amounts, but this possibility is ruled out by the fact that C_2 ensures P_2.

On the much lesser-known Interpretation 2, the evidence is not taken to support the view that we are confronted here with backward causation. PR is not assumed to have direct access to the final choices; he does not foresee them; he is only a diagnostician of the tendencies of the various CH's. His enormous success in the past in guessing correctly the final choices is explained as due to the extremely high correlation between T_i (= the tendency to do C_i) and C_i. However, as long as CH has T_1 at t_0—the time at which PR seals the opaque box—then he can be assured that P_1. Should it come about that by a strong exertion of willpower CH manages nevertheless to do C_2 at t_1, this should do him only good and he should gain the contents of both boxes.

We should pause here and note a point of paramount significance, namely, that on Interpretation 2, in spite of the great powers of the PR, he is incapable of encroaching on the freedom of CH. The latter need not be concerned with the fact that PR has succeeded in penalizing in the past all those who have chosen both boxes; he can be at complete ease and do C_2 without fear that he may increase his chances of losing \$M.

To see this, let us imagine that drug D, which formerly was prescribed for infants, has recently been discontinued because it has been overwhelmingly confirmed that any person who took the drug in infancy will lose all his hair and teeth between the ages 23 and 25. In addition, D also causes those who have been administered it in infancy to drink large quantities of coffee and never to drink tea. Now suppose that Smith, who is a teenager, does not know and is unable to find out whether or not he had been given D in infancy. He drinks coffee and loathes tea. Would there be any point in his restraining himself and, contrary to his natural tendency, to avoid coffee and force himself to drink tea? Obviously not. But experience has shown that anyone who drinks excessive amounts of coffee and never touches tea is most likely to have taken D when a baby, in which case he will lose all his hair and teeth before the age of 25; whereas if he drinks tea but not coffee, he may rest assured that he had not been administered D and will retain his hair and teeth. In spite of this we do not say that by deciding to force himself not to drink coffee but tea he can bring it about in his teens that he had not been given D a decade earlier. It is D which is assumed to cause the tendency to drink coffee and not tea, and not the other way round. All he can do now, if he can do it at all, by abstaining from coffee and drinking tea, is that if he had been given D he acts contrary to the tendency that has caused him his existing preferences. In the same way it may be said that if P_1 then the player has a tendency to do C_1 and if in spite of this he manages to do C_2 that would not bring about P_2; all it would do is cause him to act contrary to the tendency he has.

II.

I was glad to see that J. L. Mackie, without having read my book or my article in the *Brit. J. for the Phil. Sci.* 1976 in which I first made the point that there are essentially two interpretations, seems to advance essentially the same idea in his 'Newcomb's Paradox and the Direction of Causation' (*Canadian J. of Phil.* 1977). He commits a serious blunder, however, by stating that Interpretation 1 (on which we are faced with a case of backward causation — we shall explain later why) is as plausible as Interpretation 2. In fact, however, Interpretation 1 must be ruled out as impossible. The reason is simple. On Interpretation 1 the inductive evidence is taken as supporting the conclusion that C_1 is the better choice. However, there is also an argument leading to the contrary conclusion that C_2 is the better choice. The argument is fundamentally different from the (invalid) Dominance Argument: The Predictor himself is, during the interval $t_0 - t_1$, in a superior position to know which is the better choice. We may call him a Perfect Judge (PJ) to assess whether C_2 is preferable to C_1. But, assuming that once the opaque box is sealed its contents remain the same, we are in a fortunate situation to be able to work out with absolute certainty what opinion PR holds. If P_2, then surely he thinks that C_2 is the better choice, for CH will then get at least the smaller amount of money; but if P_1, likewise it is better to do C_2 and obtain both amounts. Thus he holds that C_2 is the better choice. But because his opinion is absolutely reliable, we have to contradict our previous conclusion that C_1 is the better choice.

This may bear superficial resemblance to the Dominance Argument, but it can easily be seen that the two are fundamentally different. According to the Dominance Argument, if P_2 is chosen it is better to do C_2, but also if P_1 then it is better to do C_2; hence, C_2 dominates over C_1 in both possible states. Now CH's being better off in both cases means that he succeeds in getting the contents of the transparent box in addition to whatever there is in the opaque box. And it is admitted that this may happen either on P_1 or on P_2. But because we now assume that backward causation is at work, by doing C_2 CH ensures that P_2 rather than P_1 is the case, that is, that he is better off relative to P_2. It so happens, however, that ultimately it is preferable to be worse off relative to P_1 (i. e., not getting the contents of the transparent box but getting the contents of the opaque box, which now contains the larger amount of money) than being better off relative to P_2. So following the Dominance Argument, CH really places himself in a less favorable position by doing C_2 rather than C_1. This same objection cannot be repeated in the context of the Perfect Judge argument. PR is making his judgment after t_0, when he is certain about the contents of the sealed box and he is capable of asking himself the very question: would CH perhaps ultimately place himself in a less favorable position by accepting my opinion and consequently doing C_2 rather than C_1? He knows that by backward causation the Player cannot *change* the contents of the sealed box to having been different from what he now knows them actually to be (for backward causation, of course, does not imply the changing of the past to be different from what it actually was, only the determining of what the past was). Hence, PR knows always that C_2 is the genuinely preferable choice.

This contradiction cannot be avoided by maintaining that a person cannot advise another to do something he is unable to do; and thus the so-called perfect judge cannot advise the player to do C_2 in case it is true that P_1 given the impossibility of P_1 and C_2 (cf. R. Grandy 'What The Well-Wisher Did Not Know' *Australasian J. Phil.* May 1977). This move fails for it is unnecessary for the perfect judge to advise; it is sufficient for him merely to hold an opinion, and this he may legitimately do. For example, I may ask several of my friends which of the two following choices, in their opinion, would make my wife happier: if I continued in my present job or if I became a professional boxer? Each may be aware that because of my physical condition it is impossible that any amount of training should render me sufficiently fit to last one round in the ring, and hence the choice of becoming a professional boxer is not open to me. Nevertheless, friend A, believing that my wife has very great admiration for athletes, may meaningfully make the assertion 'If you were to become a professional boxer that would make your wife happier'; whereas friend B, believing that my wife abhors all forms of physical violence, would equally and meaningfully contradict A and deny the truth of that assertion. Similarly, PR may well believe that in case it is P_1, C_2 is not open to me, and he may thus correctly assert 'If the Player were to do C_2 he would be better off.'

Thus we have arrived at a contradiction, for we have concluded both that it is best to do C_1 and also that it is best to do C_2. This may be taken as a reductio ad absurdum proof that our story contained an impossible element. The most likely candidate, of course, is the assumption that a perfect predictor exists. Thus, we may take it as proven that there is no such thing as a perfect predictor. By employing the notion of mathematical expectations in the book mentioned, it was shown that for the same reason it also follows that even a slightly competent predictor cannot exist. By a 'slightly competent' predictor I mean a predictor who possesses some competence, so that the player can increase slightly his chances that it was the case that P_1 rather than P_2 by doing C_1 rather than C_2.

It should be clear, therefore, that Interpretation 1 has to be abandoned, and we are forced to accept Interpretation 2. For that interpretation leads to no contradiction and it presents no argument for doing C_1; nothing can be lost and everything is to be gained from doing C_2.

III.

So far I have shown that the existence of a competent predictor, who has direct access to the final choices of a person rather than merely having an indirect grip on them by being capable of diagnosing their tendencies, is not possible; but I have not fully explained why not. It may well be suggested that the reason is rooted in the impossibility of backward causation. Some philosophers have said that if the player's choice could be predicted then we might be having a case of backward causation (e. g., J. L. Mackie in the article referred to). The fact that backward causation can never occur may, in turn, be explained by saying that the future is unreal. It might be said that because of the unreality of the future, it is not possible for events that have not yet occurred to shape the present, as they do not exist to be able to do so.

The immediate question to be asked is: why is it believed that if the predictor were competent qua predictor then we would be facing here a case of backward causation? The answer that may be suggested is that there is an asymmetry between the predictor and the player, through which the predictor's act is dependent on the act of the player and not the other way around. The player does what he wants; he can decide, if he wishes, to do C_1 in case he wants to bring about P_1; whereas if he should freely choose to forego the larger amount of money then he may choose to do C_2. On the other hand, the predictor does not do either P_1 or P_2 according to what he freely wills; his action depends on whether C_1 or C_2 is the case. If C_1 is the case then he must do P_1; otherwise he is compelled to do P_2. It stands to reason to look upon the choice of the player, which is made according to his own wishes, as the cause, and upon the act of the predictor, which is determined by this choice, as the effect.

It seems to follow from this that there is no reason why, if the human player is substituted by a mindless machine, that its moves should not be predictable either. Given that the machine is mindless, it has no will, and hence the previous argument does not apply and therefore its moves could be predicted without thereby presenting us with a case of backward causation. But then we have a problem, because my argument that a player's choice is in principle unpredictable seems to work just as well when the player is a machine as when he is a human being.

The answer is that there is no valid proof in the case of a machine that its moves are unpredictable. In my book *Religion and Scientific Method* I have argued in great detail for this. (But apparently, the argument is difficult to grasp and/or accept.)

For example, J. B. Maund, who offers a thoughtful review of that book in the *Australasian J. Phil.* 1979, says that he found that my 'argument is hard to follow' (p. 119). So I shall attempt to offer another one which may be easier to follow:

The deterministic machine cannot on Interpretation 1 derive the relevant contradiction. Admittedly, it can summon Argument 1, based on induction, which leads to the conclusion that because in the past everyone who took the opaque box only gained more money than those who took both boxes, it is best to do C_1. It does not, however, have access to Argument 2, whereby it follows that it is best to do C_2. Let us once more look at what that argument entails. The following could be suggested:

Arg. 2 (α)   If P_2 then obviously PJ would say it is best to do C_2; hence I will do best by doing C_2 if P_2.

(β)   If P_1 then still PJ would say it is best to do C_2; hence I will do best by doing C_2 also if P_1.

(γ)   But either P_1 is the case or P_2 is the case.

$\therefore$ (δ)   I will do best by doing C_2.

This argument does not work because it has a false premiss, namely, (β). The machine cannot argue that even if P_1, it would nonetheless gain most by doing C_2, for it is clear that it does not stand to gain more (nor for that matter does it stand to gain less) if it does C_2, as it is out of the question that he should do C_2. There is overwhelming evidence that no matter how the machine argues, the end result cannot be C_2 given that P_1. In the past whenever P_1 obtained, all CH's whether people or machines were involved, they ended up doing C_1. Thus, (β) is definitely false and Argument 2 collapses.

Now it might occur to the reader that he could demolish my original reasoning by claiming that Argument 2 (on Interpretation 1) does not apply to a human selector either, because (β) is false no matter who CH is. But this would be a grave error and in its realization lies perhaps the most crucial part of our story. I shall now proceed to expose briefly, but as clearly as I can, this error.

The reader should attempt to put himself in the shoes of the CH. If he is a person who is capable of grasping the meaning of (β), and undoubtedly he is, then he must say quite categorically that he wants to do C_2. Now it might be objected 'but if P_1 has obtained, then in spite of the fact that CH may most earnestly desire to do C_2, he simply is not capable of doing so'. If so, this would abruptly change the situation, and we would no longer be dealing with a case that had anything to do with Newcomb's problem. In that problem CH was to weigh rationally the available choices and decide for himself what seemed to him the best choice, assuming that there was a point in doing so, that is, that he would be able to act on his decision. (At any rate if it were indeed the case that if P_1, then even if CH wanted to do C_2 he could not do so, and we would still have an indication that backward causation was impossible, for we would find it impossible to construct a game in which CH does C_1 or C_2 at will to bring about P_1 or P_2). So it is clear that the human selector must be permitted the freedom of doing either C_1 or C_2, depending on what he wants to do. In the case of a mindless machine, this is not a consideration, for it cannot act in accordance with its wishes (nor for that matter contrary to its wishes) as it is stipulated that it does not possess the mental capability of possessing wishes.

IV.

For some reason, the investigation of which is more a topic for psychologists than for philosophers, many people have resolutely refused to understand the central argument leading to the conclusion that Interpretation 1 harbors a contradiction. With varying ingenuity 'objections' have been invented, a large number of which I attempted to tackle in *Rel. & Sc. Meth.* But new objections keep cropping up. The following is by André Galois in *Analysis* 1979.[4]

In an effort to discredit my PJ, Galois constructs a somewhat different game, Game II, in which instead of the PR we have an Observer (O) who places the two boxes in front of the CH but leaves the opaque box empty while he puts $1000 in the transparent box. The CH is still allowed either C_1 or C_2. It is also given that after the CH makes his choice, O finds out what the choice was and

(A')   If O observes that CH did C_1 then he puts $M in the opaque box (which the CH may pocket).

(B')   If O observes that CH did C_2 then he puts nothing in the opaque box.

Now Galois claims—and nobody in his right mind would disagree with him—that here it is absolutely certain that it is in the CH's best interest to do C_1. Consequently, he should not attempt to do C_2, which would result in getting no more than $1000. Yet, Galois argues, it seems that here just as in Game I there is a sound

4. "How Not to Make a Newcomb's Choice" (pp. 49-53).

argument for maximizing profit by deciding to do C_2: when CH had already made his choice whereupon O did or did not put $M in Box I in accordance with A' & B', O may reason that if now PJ inspects Box I and

> . . . discovers that Box One[5] is empty, he will want me to have both boxes, for I will then end up with $1000 instead of nothing. If he discovers that Box One[5] contains $M he will want me to have chosen both boxes, for then I will end up with $1,001,000.[6]

Obviously, Galois claims that we would all agree that it would be a mistake to accept the opinion of the PJ according to which it is never to the disadvantage of CH to decide to do C_2. Similarly, he says my argument in the context of Game I is also mistaken.

Contrary to Galois, however, there is a crucial difference between the situations that obtain in the two games. In Game I, where PJ is consulted before CH makes his choice, we may put to him the question: is it possible for CH to lose anything should he attempt to do C_2? I am certain that Galois would not wish to deny that even if PJ knows that Box I contains $M, he will insist that no harm can come to CH by making an attempt, however determined, to do C_2. He might of course point out that given that the opaque box is not empty, it is virtually certain that his attempt will fail in view of the extraordinary record of PR. Should he, however, manage somehow to do C_2, then for the first time in history a CH would succeed in breaking the link between the actions of the PR and CH. Even then, of course, the CH would not sustain a loss but on the contrary gain an extra $1000. Under no circumstances can be fail to acquire the $M once it is given that it is in Box I.

Suppose, however, that in Game II PJ sees $M in Box I, in which case he learns that CH has done C_1. It is definitely not the case that the PJ would maintain, that given that there is $M in Box I, CH could not have lost anything had he attempted to do C_2. For if CH had tried to do C_2, there is no reason why he should not have succeeded, and consequently there would be no money in Box I, because O would not have put any there. There is absolutely no reason to say that Box I would still contain $M, even if earlier CH would have done C_2.

Consequently, we may now state for the sake of greater clarity the contradiction that arises in connection with Game I in the

5. [i. e., the opaque box.]

6. ibid., p. 51.

following manner. Suppose we ask, is it possible for CH to lose anything by attempting to do C_2?

Argument 1:   Given the successful past record of PR, which shows that without exception all those who had done C_2 ended up with a mere $1000 while all those who had done C_1 received $M, the answer is yes.

Argument 2:   The answer based on the assessment of PJ as shown before is no.

Thus, we have a contradiction which forces us to abandon Interpretation 1.

To conclude with an illustration of what confusion professional philosophers are capable of I shall cite R. G. Swinburne who in *Philosophical Books,* 1979, p. 86 has this to say:

> . . . I cannot see why the predictor cannot observe the state of the brain, and infer the wants to which it will give rise, the arguments which the human agent will use to himself, the conclusion which he will reach and whether he will act on his conclusion. Oddly, Schlesinger allows (p. 106) that there might be a "perfect diagnostician of tendencies", e. g. of a tendency to choose both boxes; and Schlesinger gives no argument to show that the tendencies might not be deterministic ones. But he seems to think that the diagnostician is not a predictor. Again, he writes (pp. 111f.) "freely willed choices are inaccessible to prior discernment, although they can indirectly be guessed with near certainty via the tendencies imprinted in the brains of the agents"! (Again no argument is given why the certainty must only be "near". He thinks that an agent might act against his tendency, but he gives no argument to show that he can.) It is difficult to know what to make of all this. Schlesinger seems to be moving from arguing that neither the agent's choices would be better justified than the other to arguing that the choice is not predictable. But if this is what is going on, clearly the step is fallacious.

I wish to cast no aspersions on Professor Swinburne's sincerity. I am willing to believe that he is as totally uncomprehending as he claims to be. Yet I also want to believe that most people will have done better than he. Most people, for instance, will have been able to 'see why the predictor cannot observe the state of the brain, and infer etc.', by realizing that if the predictor could do this then he would have direct access to CH's final choices; and CH would ensure that by doing C_1 he gains more than by doing C_2, thus underwriting Argument 1 by which C_1 is preferable to C_2. However, this is contradicted by Argument 2 from PJ, which leads to the opposite conclusion; and we cannot entertain assumptions that lead to a contradiction.

As to Swinburne's last two sentences, the thoughtful reader will realize at once that having Argument 1 and Argument 2 does not simply imply that neither C_1 nor C_2 is the best choice but rather that because we have ended up with the conclusion that C_1 and not C_2 is the best choice, as well with the conclusion that C_2 and not C_1 is the best choice, we have a contradiction which compels us to abandon the assumption that the predictor is competent qua predictor. This, then, is what leads us to conclude that the choices are not predictable. (v. Q&A 48–52)

*   *   *   *   *

Do we thus have a conclusive proof that backward causation is in principle impossible? Not quite. The fact that Interpretation 1 had to be renounced is well explained by suggesting the impossibility of backward causation being its source. The hypothesis that backward causation can never occur is a good explanation and, to that extent, the hypothesis is confirmed by our findings. But it is certainly not the only explanation and it is very hard to decide whether it is the best. To name one other explanation, one which was discussed in *Rel. & Sc. Meth.*, Interpretation 1 is out of the question because it would entail the lack of genuine freedom of choice.

8 *Conventional Events*

We shall look at one more attempt to show that backward causation is in principle impossible. This will involve events of the type of causally connected event-pairs in whose case it is entirely obvious which event is the cause and which is the effect. I am referring to cases where one of the pair is not a physical but rather a conventional event. For example, the signing of a certain document is causally connected with the legal event of a certain item passing from the possession of the seller into the possession of the buyer. Everyone would agree that it is the signing of the document which brings about the sale rather than vice versa. Similarly, it is the wedding ceremony that changes the marital status of the two people concerned rather than the event of a man and a woman turning into a husband and wife which causes the wedding ceremony.

Of course, the conventional event in question need not be a legal event. Take for example, a game of chess which is being won by one of the players making a certain move. In this case it would not occur to anyone to claim that perhaps it is the winning of the game which causes the move in question.

There is unquestionably a fundamental difference between physical and conventional events. Yet many important aspects of the connection between, say, the signing of a contract and the sale of a piece of property are identical to aspects of the connection between ordinary causally linked physical events. There is a regular conjunction between instances of the closing of a sale and of property transference. The sale may be said to be generated, brought about, or determined by the signing of the document of transfer. As causal laws concerning physical events support counterfactuals, the law concerning the transaction of properties supports the counterfactual 'Had I signed that document my house would have become yours'. Thus, the study of the causation of conventional events can be of great use for our purposes.

First of all it becomes once more evident that precedence in time is not a conceptually essential feature of a cause without which the cause could not be distinguished from its effect. In the cases before us, the events in question normally occur simultaneously, and yet we immediately identify the cause and the effect.[7]

Another instructive feature of these cases is that they seem to confirm what we have said in a previous section—namely, that an asymmetry between a cause and its effect is manifested in the different ways in which an interference influences the cause and its effect.

Let us consider a situation in which someone wishes to sell his property to someone else, and, after negotiations, comes to an agreement. Finally, in front of their lawyers the parties put their signature to a document which is to close the sale. Let us call this situation Situation 1.

---

7. It is interesting to note that this point can also be gotten hold of back to front. J. Kim in his 'Noncausal Connections', *NOUS,* 1974, pp. 41–52, considers Xantippe's becoming a widow as a result of Socrates' death. He argues that the connection between these two events cannot be a causal one because they are simultaneous. Although the aspects mentioned in the above paragraph can be claimed to necessarily belong to a causal connection, I do not see why non-simultaneity should be looked on so. He advances an even weaker argument why such connections cannot be causal:

"... if it is plausible to locate these events at different spatial positions we would have to accept this case as one in which causal action is propagated instantaneously through spatial distance" (p. 42).

Situation 2 is identical with Situation 1 in every respect except in one. In Situation 2, a tiny bit of dirt entering into the opening of the ball-point pen used for the signing of the document blocks the flow of ink, and, although the parties go through exactly the same motions of signing as in Situation 1, the place of the signatures remains entirely blank. The blockage occurring in the ball-point pen constitutes an interfering event which prevents the event of signature taking place; consequently, the property fails to pass from the possession of the seller into the possession of the buyer.

In Situation 3, the world is exactly the same as in Situation 1, with the exception that the proper authorities issue a decree that at a certain date no transaction of properties may take effect. It is clear that this piece of legislation constitutes an interference with the event of the sale, which now cannot take place in spite of the fact that the document is fully signed just as in Situation 1.

Thus, in Situation 2, the occlusion in the ball-point pen which prevents the signature from being recorded also interferes with the normal effect of such a signature. Whereas in Situation 3, the decree which prevents the sale from occurring has no effect on the cause of the sale, the signature. This difference between Situations 2 and 3 seems to me to show clearly, as we have said in a previous section, that whenever A is a sufficient and necessary cause of B in the absence of any interference with either of them, then the occurrence of an event, which it is proper to describe as an interference with A, automatically brings about the prevention of B as well; whereas a direct interference with B (preventing its occurrence) may permit A to occur nevertheless.

At the same time, it is obvious also that the causation of conventional events has features which are not duplicated in the causation of physical events. Indeed, the very feature that is responsible for distinguishing the cause from the effect in the conventional case is unique. The reason why in the conventional case we are so certain which is the cause and which is the effect is the obvious total

---

So what of it? Special relativity of course teaches us that physical disturbances, for example, electromagnetic waves, cannot travel at a speed faster than light, but nothing about the propagation of causal action of the kind discussed here or by Kim.

Incidentally, I keep away from examples such as Kim's, because it is not clear that, for instance, Xantippe's widowing is a separate event from Socrates' death. Kim's arguments to the effect are unconvincing. For example, he says that the location of Xantippe's disespousal is not in the prison where Socrates dies. But what about Xantippe losing a husband? Where did that happen?

There is no doubt however that the signing of a contract and the transference of a piece of property are separate events.

dependence of the effect on the cause. This is so because the effect is defined in terms of the cause; it has no independent existence at all. We cannot directly observe a transaction taking effect, for our knowledge of it comes about through our knowledge of the physically observable events which bring forth the transaction. We can certainly not duplicate among physical events the unique relationship of complete dependence between cause and effect which we find in the causation of conventional events for it is an essential feature of the effect that it is not a physically real event.

It is also clear that the question whether or not nature permits backward causation cannot be decided by observing whether the causation of conventional events happens to be permitted to occur in the reverse order in which the effect precedes its cause. Nature may never permit an effect to precede its cause. However, this is not the reason why an event whose existence is secured by human convention should not be postulated as having occurred before the occurrence of its physical cause. Thus, the constitutionality of retroactive legislation of parliament, or the sanctioning of a law permitting that a wedding ceremony today should confer upon a man and a woman a married status as of twelve months ago, would not indicate of course that among physical events, too, backward causation might be possible. Nor would the contrary fact, that such cases of backward causation were never permitted in human societies, teach us anything concerning physical events. Yet as we shall see, the study of the causation of conventional events seems to be able to help us in our inquiries concerning the possibility of the backward causation of physical events.

9   *The King, the Nobleman, and the Serf*

If the future is unreal, it may be so contingently or necessarily. It may be only a contingent fact that the events of the future have no reality at present. But it may also be built into the very notion of an event not yet materialized that it cannot be real now. In the latter case, backward causation may not only be forbidden by nature but also ruled out by logic in the same sense, for example, that it is logically impossible for a nonexisting object to occupy a given point in space. I take it that most people would agree that what does not exist cannot have a position in space, for built into the notion of physical space is the concept that anything physically nonexistent can have no position in it. Similarly, it might be argued that it is logically impossible for something that does not yet exist to act and bring about anything. If this is indeed so and if we, nevertheless, postulate that a yet nonexistent event has a causal impact on the present state of affairs, this will of necessity ultimately give rise to a manifest contradiction. Therefore it seems that, if we find that in a situation in which we have postulated the occurrence of backward causation we are involved in a contradiction, we then can reasonably conclude that this is an inevitable result of the fact that we have postulated the occurrence of something that was logically impossible. With this in mind, we shall have a further look at event-pairs in which one of the events is a conventional one.

We shall consider the story of a medieval king who, on January 1, 1234, decrees that a serf be released immediately from the bondage of the nobleman to whom he belongs, on the condition that for the coming year the serf transgresses no law of the realm, voluntarily or involuntarily. According to the king's decree, the status of the serf during the whole year of 1234 is to be unkown; but at midnight, January 1, 1235, if it turns out that he has been completely law-abiding during the preceding year, we will then accept the claim that he has been a free person throughout the whole year. Otherwise, he has been and continues to be a serf. On June 1, 1234 the serf sells his horse to someone, and the nobleman, on hearing about it cancels the sale, using the privilege given to him by law to cancel at will any legal transaction of his serf. On the next day, the serf works with the horse without asking permission of the man to whom he sold it the day before. Otherwise, the serf is meticulously law-abiding throughout the entire year.

Is he to be declared on January 1, 1235 a freeman since January 1, 1234? There is an obvious difficulty here. If we say that he has been free the entire preceding year, then the nobleman had no right to cancel the sale of his horse. And thus on June 2 the serf was working without permission with someone else's horse, thereby transgressing a law. Consequently, he is not free. But if he is not free, then the nobleman did have the right to cancel the sale and, therefore, on June 2 the serf was working with his own horse, not transgressing any laws and thus became free from January 1, 1234. This is a contradiction.

It may seem to some that there is no fundamental problem here. What we need to know is simply the laws applicable to a person of a special status in which our serf finds himself during the year 1234. Is he to obey the laws pertaining to serfs during this year in which he is, so to speak, in limbo? If so, the nobleman had the full right to cancel the sale and there is no problem. Or, on the other hand, is he to obey the laws that all other free persons must obey, and is his legal status in every respect that of a freeman? In that case the nobleman could not cancel the sale and there is no problem either. It is also possible that because of his halfway status, he is to be regarded partially free and partially a serf, thus making him subject to some of the laws of serfs and some of the laws of free people. In this case, we have to inquire whether the particular law conferring on the nobleman the power to cancel the transactions of his serfs does or does not apply to our serf. If nothing can be found in the lawbooks concerning this case, then it is clear that the system of laws in the realm is defective, for is does not cover certain cases which may occur, and we should request the King to issue a decree covering this case.

It seems to me, however, that this is not the right approach. For what if the king has resigned in the meantime, nobody in the meantime has the authority to make new laws, and a judge must decide the status of the serf? More importantly, the king may well insist that his laws cover every contingency, and that there are no individuals who are partially free and partially serfs. In our case, we may not know what the individual in question is, except that he is fully a serf or fully free, and therefore either all the laws of serfs or all the laws of free people apply to him. Before all the facts bearing on his status are in, we may not know which is the case. By January 1, 1235, every relevant bit of information is known and if there is still a difficulty it is not because some legislation or evidence is missing but rather because something is essentially wrong with our story. At any rate it is not significant that even after the year ends we still do not know whether the serf is free or not. What matters most is that we have positive proof that if he is free then he is not and if he is not free then he is. Thus we have 'p if and only not-p' which is a contradiction. By reductio, something in our story is logically impossible. It would seem that the most likely candidate is the king's decree which postulates that later events, i.e., the behavior of the serf during the year 1234, in particular that during all this time he refrain from breaking the law, should determine his assuming or failing to assume a new legal status earlier, namely, on January 1, 1234. The contradiction we have arrived at should thus be taken—it might now be suggested—as indicative that we have begun with a story that harbors a logical impossibility, namely, the postulation of backward causation. It follows then that it is logically impossible that the occurrence of an event in the present should be dependent on another event which has not yet occurred.

Unfortunately, however, there is no conclusive proof that this is so, for it may be that less than the wholesale rejection of backward causation is what is in fact required. It could be objected that in general backward causation might be possible, and the only objectionable feature would be for causes and effects to form a closed circle. This is, if S_1 is a state that causally affects a later state S_2, then S_2 must not in turn affect S_1 itself in a way in which it would modify the very influence S_1 is to have on S_2. In our case, however, this is precisely what happened. We may look on the serf's state on January 1 as S_1 and on this state on June 2 as S_2. Now S_1 has a causal influence on S_2, for if the serf is free on January 1, then his working with the horse on June 2 is an illegal act. But now S_2 is supposed to have a causal influence on S_1 and modify it, for if the serf's act was illegal, he is not a free person on January 1, and perhaps this is what is wrong with our story.

We end this chapter with the conclusion that backward causation is a phenomenon that in principle is discernible, but whether it may or may not occur requires further investigation. (v. Q&A 53)

V

THE STATUS OF THE PAST

1 *Scepticism Concerning the Existence of the Past*

In the last chapter we discussed the question of the status of the future; in this chapter we shall discuss the question the status of the past. A particular form of scepticism exists concerning the existence of the past which deserves special attention, because it differs from any other form of scepticism. The sceptical position to which I am referring was advanced by Bertrand Russell: it consists in wondering whether the universe is more than five minutes old. Let us call the normally held hypothesis according to which the universe is many billions of years old H_1, whereas H_2 shall denote the hypothesis according to which the universe sprang into existence just this moment (as we may well be sceptical about the reliability of our memories, we might as well not allow the existence of a past of even five minutes duration), with all the traces of an old universe. On H_2 the future history of the universe is not expected to be different from what it is supposed to have been on H_1. There is thus no evidence we may hope to encounter which could adjudicate between H_1 and H_2. It may also be supposed that neither of them has a marked feature which would make it intrinsically preferable to its rival.

The point at which I shall begin is an inquiry as to whether H_2 is an 'interesting' hypothesis. Russell himself did not think so, although he didn't explain why not. Normal Malcolm, on the other hand, thought that although H_2 was interesting it was logically impossible.[1] James W. Cornman, I believe, has satisfactorily proven that H_2 is logically possible but he seems to revert to the view that it is quite an uninteresting hypothesis. He hints at an explanation for this by suggesting that H_2 is not interesting because 'there is no reason to consider seriously such a possibility'.[2] This I find puzzl-

1. "Memory of the past," *The Monist,* 1963.

2. "Malcolm's Mistaken Memory," *Analysis,* 1965, p. 167.

ing. Why should we consider such a possibility less seriously than other possibilities which lead to other forms of scepticism? Does he possess some strong arguments, which he has kept to himself, why scepticism concerning the existence of the past, unlike all other forms of scepticism, is devoid of serious interest?

I venture to suggest that the actual reason why Russell regarded as uninteresting the sceptical position he invented is as follows: all the other known sceptical positions may be considered interesting because they have the potential to affect the attitude of anyone who entertains them. For example, a person who is truly unconvinced that an external world exists might be tempted to throw off all ethical constraints. As other people are figments of his imagination, there is no reason why he should nurture feelings of compassion. Thus, there is an incentive to look for ways of combatting scepticism concerning the existence of an external world, if for nothing else than to challenge any philosophical basis for callousness. A person doubting the existence of other minds might similarly feel free of all ethical restrictions. If others may be mere robots, lacking sensations, emotions and awareness, there is no reason to feel compassion toward them. So once more scepticism concerning other minds is 'interesting', for it has practical implications which are worthy of being combatted. Or consider scepticism concerning the validity of induction. He who does not believe that the inductive method produces reliable results may not step out of the way of an oncoming car, may walk beyond the edge of the roof, or do all sorts of things considered dangerous by others on the basis of past experience. Scepticism concerning the validity of induction is 'interesting', and it makes good sense to try and prove it so as to show that the results of empirical science are reliable.

Scepticism concerning the reality of the past seems to be very different. Doubting the truth of H_2, or even firmly upholding the truth of H_2, would not seem to provide any reason for adopting an attitude different from one held by someone else who was convinced of the truth of H_1. After all, H_1 and H_2 seem to have exactly the same implications with regard to all aspects of the world for any time yet to come. H_2 therefore need not concern us. Let anyone adopt it; it is an utterly harmless hypothesis.

The point I wish to make is that this view is false. If we were really in doubt with regard to the reality of the past, our conduct would of necessity differ from that which we practice under present circumstances. For it is not true that it is of no consequence whether we subscribe to H_1 or H_2. Suppose that there is overwhelming evidence provided by numerous eyewitnesses that one week previously A has committed a crime. According to H_2 there is no basis to hold that A has actually committed a crime a week ago because the universe did not exist a week ago; but according to H_1, A's guilt may be regarded as firmly established. Are we justified in punishing him? According to some theories of punishment, it may not matter that perhaps A did not commit the crime, for the mere fact that everyone believes he has committed a crime renders punishing him a useful service for society. This is a theory to which many philosophers would not subscribe. For them the only way in which to justify punishment is by pointing out that it is just retribution: a criminal by his act has rendered himself deserving of punishment. But on the retribution view of punishment, H_2 entails that A not be punished because at the time A allegedly committed a crime, the universe did not even exist; whereas H_1 entails that he should be punished. But if we have no proof that H_2 is false, on what basis do we in general punish criminals?

It may be pointed out that when faced with a situation in which there are two competing hypotheses, H_1 and H_2, and there is no possibility of producing evidence in favor of either of them, philosophers disagree regarding the correct way in which to regard the dispute. Philosophers who formerly subscribed to the verificationist theory of meaning maintained that because the question of whether H_1 and H_2 is true does not lend itself to being determined by observation, the question itself is devoid of meaning. In other words it is meaningless to inquire whether H_1 and H_2 is true. Nonverificationists, on the other hand, maintain that we merely have no way of knowing whether H_1 or H_2 is true.

If we leave matters as they are and do not show why Russell's scepticism is unjustified, we are in difficulty no matter which philosophical position we adopt. After all, it is not possible to maintain either that it makes no difference whether A is punished or that the question whether he should be punished is meaningless. Nevertheless, criminals *are* punished; and on the surface this fact seems to present a problem both to the nonverificationist and to the verificationist. To the nonverificationist: because he has no evidence by which to resolve the question whether H_1 or H_2 is credible, why in practice do we seem to assume that H_1 is true? To the verificationist: if he is to insist that the difference between H_1 and H_2 is illusory, how does that explain why we act as if we took H_1 rather than H_2 to be true?

In what follows I shall discuss the verification principle, show how the principle may be clearly stated and then show how it may be employed to solve the problem we have raised. (v. Q&A 54 and 55)

2 *The Verification Principle*

Verificationism is one of the most exciting ideas produced by twentieth-century philosophers. Yet it has been widely abandoned, and even discussions of it are rare. The facts leading to disillusionment with verificationism have been documented in detail many times. Briefly, philosophers have forsaken verificationism because after prolonged efforts to formulate a satisfactory verifiability principle, or a more liberal confirmability principle, they have concluded that such a formulation is beyond reach. But if the principle cannot even be stated clearly, it appears unreasonable to demand that all meaningful empirical sentences comply with it.

In this section I intend to show that there is no problem in stating clearly the confirmability principle and to present a simple statement of the principle. There is, however, a problem, as I shall point out, regarding the use to which verificationism is to be put. I shall suggest a novel way in which it may be applied to the allaying of scepticism.

First then to the statement of the principle. It is hard to escape a sense of amazement at the perplexity of those who concluded that verificationism is unintelligible because its meaning cannot be clearly stated. After all, a simple and entirely unobjectionable rendering of the confirmability principle may be suggested as follows:

> A sentence is confirmable in principle if, and only if, a situation is coherently describable in which it would have to be regarded by the tenets of scientific methodology as confirmed.[3]

Thus, if we come across a sentence which, in the prevailing situation is confirmed neither as true or false, we need to see whether we can coherently describe a situation in which the sentence would have to be regarded as having been confirmed. If we can describe a state of affairs in which certain features of the universe would have to be construed as evidence supporting the truth or falsity of the sentence, then the sentence would have to be regarded as confirmable in principle.

It might be objected that we have only defined 'confirmable' in terms of what is 'confirmed', leaving the latter notion undefined. The objection is groundless, however, because the notion of 'being confirmed' is not generally regarded as vacuous. It has not been

---

3. It should be understood that 'evidence E confirms the truth of S' means no more than 'E renders the credibility of S higher than it would be in the absence of E'.

maintained by many that it ought to be abandoned as useless. True enough, some have called for the rejection of the concept of confirmation, e.g., Karl Popper. But even those have some parallel notion (as for instance Popper's 'corroboration') which is supposed to contribute—like confirmation—to the acceptability of a hypothesis. At any rate it is clear that the great majority of philosophers do not look on the notion of confirmability as inexplicable and unintelligible. In fact, it is fairly commonly agreed that to make a sentence credible it must be positively confirmed to a higher degree than any of its rivals. Admittedly, philosophers have not yet been able to describe the general conditions sufficient or necessary for a hypothesis to be confirmed by given evidence. But this fact has served only as an incentive to invest efforts in the ongoing enterprise of developing a 'theory of confirmation'. It has not led to the universal rejection of the notion of confirmability as intrinsically obscure.

Consider, therefore, the case of a sentence S which, in the prevailing circumstances, is neither confirmed to any extent as true or false nor are we capable of describing any situation in which it could be regarded as being confirmed. Surely then we would say that we are in doubt whether S is confirmable in principle. There would be no justification for concluding that the notion of 'confirmability' is essentially obscure and should not be the notion to which meaningfulness is tied. After all, to the extent that there is unclarity concerning 'confirmability', there is also a doubt concerning 'confirmed'. No one has suggested that credibility must not be linked with being confirmed, nor claimed that the notion of confirmation is intrinsically unintelligible and therefore we do not know what is required when it is said that for a sentence to be made credible it must be confirmed as true to a higher degree than any of its rivals.

The situation appears even more remarkable when we note that the required relationship between meaningfulness and confirmability is much simpler than between credibility and sufficient confirmation. For a sentence to be meaningful it is enough that there be the possibility that some evidence would confirm it; the complicated question of degrees of confirmation does not arise at all. On the other hand, before S can be accepted as credible, it has to be shown that it is confirmed to a higher degree than any of its rivals. And yet because of the relatively simple problem that has to be solved in order to decide whether S is meaningful, namely, whether S is confirmable to any degree, meaningfulness when tied to confirmability has been declared too obscure a notion. On the other hand, the much more complicated problem of comparative degrees of confirmation, which has to be solved before we know whether S is to be accepted as credible, has not been seen as making the notion of credibility obscure!

It may be worthwhile noting that the verification principle which, in essence, is a principle about the meaningfulness or significance of sentences, may be stated without the term 'confirmation' featuring in it at all:

> A sentence is empirically significant (or meaningful) if and only if circumstances are coherently describable under which its credibility or acceptability would be higher than in the absence of these circumstances.

If someone were to ask, what precisely are the circumstances under which the credibility of a sentence rises? he would have to be directed to the vast amount of work done on this topic by philosophers of science. The bulk of the work done by these scholars is about the nature of observational or evidential support, theory construction on the basis of this support and the acceptability or credibility of theories and hypotheses. In studying the history of science, interest focuses on the changing fortunes of the different hypotheses and theories which, after gaining credibility, are eventually modified or entirely replaced when other hypotheses and theories acquire credibility at their expense. Although it may be a long time before all the intricacies of the notion of empirical support contributing to the credibility of scientific assertions are laid bare, its legitimacy requires no defense.

If something does require further inquiry it is in fact the problem of how it is that 'confirmability in principle', which is so immediately definable in terms of the possibility of empirical support, has for so many years been considered inexplicable. The solution emerges when we remember that originally 'verifiable in principle' was intended to mean something much stronger than the mere possibility of a sentence having its scientific credibility raised, something nearer to conclusive verification. Hence the notion was not recognized as related to the concept of scientific confirmation and was thought of as requiring definition from scratch. As years went by, the notion of 'confirmability in principle' continued to weaken and eventually no more was meant by it than the possibility of the standard kind of support hypotheses receive in science. Now, however, the construction of one verification criterion after the other has become so much a part of what verificationists are supposed to do that it no longer occurs to philosophers that an independent definition of 'confirmability in principle' is no longer required. The considerable progress philosophers had been making in constructing a confirmation theory contributed to the clarification of 'confirmability in principle' no less than that of 'confirmed in practice'. (v. Q&A 56, 57, 58, and 59)

3 The Problem of the Use of the Verification Principle

The problem that deserves our continued interest is this: what function is there for verificationism to perform; that is, what useful results, if any, does it yield?

When verificationism burst upon the world, its adherents envisaged that it would revolutionize philosophy by showing that some of its most vexing problems, which had worried philosophers for centuries, were pseudo-problems. These problems concerned the question of which metaphysical theory should be adopted in various instances, but all metaphysical disputes were declared devoid of meaning by verificationists, as they could not be adjudicated by any possible experience. This conclusion, however, proved to be far too hasty. Although it is characteristic of metaphysical controversies that their rival theories, unlike rival theories in science, do not yield different predictions, they are nevertheless controversies about the correct amount of some undisputed features of the universe. When a philosopher maintains that T_1, and not T_2, is the correct metaphysical theory to explain certain aspects of the world, some of the ways in which he may argue for this position include either the claim that T_2 is inconsistent, or that it does not fully account for every component of the relevant phenomena, or that it contains redundant elements or it is in some other way logically inferior to T_1. If his arguments are valid, then his contention that T_2 must be rejected is certainly legitimate and not meaningless.

The verifiability principle, however, may have its application in cases where we have two theories T_1 and T_2, both of which account fully for every feature of our observations; where no prediction follows from one that does not follow from the other; and where we can find no feature that distinguishes T_1 from T_2 which could serve plausibly as a basis on which to prefer one to the other. In such a situation, the difference between a verificationist and a nonverificationist should emerge. The latter would have to conclude that we have no method to find out which of the two is true. The verificationist's position will be basically different. He will say that because nothing exists that points more to the credibility of T_1 than to T_2, and because we cannot conceive of circumstances under which there would be anything so indicating, there is no difference of meaning between T_1 and T_2. On the surface T_1 may look different from T_2, but after subjecting the matter to the verificationist test, it becomes evident that the difference is a matter of wording and not of substance; the question whether in fact T_1 is true or T_2 is true makes no sense.

Some philosophers, while agreeing that we have described a concrete difference in the attitudes of a verificationist and a nonverificationist, may find the discussion altogether unexciting. They would claim, not unreasonably, that if this were all there was to it, verificationism would have shrunk enormously in scope from what originally was assigned to it.

There is thus a call for an exploration of the possible uses of verificationism. A sketch of an entirely novel line of approach is given in what follows.

4 *Allaying Scepticism Concerning the Existence of the Past*

We shall consider a suggestion a verificationist (V) may advance in order to allay the Russellian scepticism concerning the reality of the past. V begins by observing that many philosophers have agreed that I may assert incorrigibly 'I remember that p' where p stands for 'yesterday it rained'. This is so when 'remember' is not used to function like 'know', in which case I could not be said truly to remember something that did not actually take place, but where 'I remember' functions merely to report that a given impression prevails in my memory. The statement then is an avowal of my immediately sensed experience, on which I am according to many philosophers an absolute authority. It would make no sense to question that perhaps it only *appears* to me that I remember p, when there is no distinction between 'It appears to me that I remember that p' and 'I remember that p'.

V would continue by claiming that if 'I remember that p' is true then p must be meaningful. According to V's claim 'I remember that p' lies between 'I know that p' and 'I believe that p'. The former, if true, entails that 'p' is true; the latter, if true, does not even entail that 'p' is meaningful, for I may believe that p because I was told so. It is to be understood that when I say that I remember that p, I mean to assert that I directly remember that p (i.e., I retrospectively remember p, to use Furlong's expression) and not, for instance, that I heard reliable reports that p. The statement then is an avowal of my immediately felt experience, on which I am according to many philosophers an absolute authority. It would be absurd for anyone to attempt to talk me out of it, to try and convince me that I am mistaken and that it is not a fact that I remember that p.

It is essential to emphasize that V would not regard 'I remember that p' incorrigibly true irrespective what p stood for. When p stands for 'yesterday it rained', then the statement is incorrigibly true because there are two necessary factors present: (i) The kind of sensation I claim to be undergoing is certainly such that anyone else can at will reproduce it in his mind. You know precisely how it feels to have the memory impression that it rained yesterday. I am thus asserting that I am having an experience of a kind with which you are perfectly familiar and therefore there is not a shadow of doubt that you know what I am talking about. (ii) Being mistaken is out of the question because I am talking about my immediately sensed experience, and being untruthful is ruled out as it is given that I am sincere.

However, when 'p' stands for something such as 'the Absolute was lazy', then 'I remember that p' is not true because condition (i) is not fulfilled. When I claim to remember that the Absolute was lazy, you, the observer, really have no idea what kind of a sensation I am reporting to be undergoing when allegedly experiencing the remembering of what I claim to be remembering. In fact, you have no idea at all what I am talking about.

Thus, we may well suppose that V would be claiming that in all those cases in which (i) and (ii) are fulfilled, and consequently 'I remember that p' must be accepted as true, then 'p' is meaningful. He is not claiming, of course, that if 'I remember that p' is true then p is also true; only that p must then be meaningful. V would argue that 'I remember that the Absolute was lazy' is not true in spite of my being an unquestioned authority of any immediate sense experience, simply because 'the Absolute is lazy' is non-sensical and nobody can imagine what sort of sensation I must be partaking in when perceiving the Absolute to have been lazy. Thus, if 'p' is not meaningful, then neither is 'I remember that p'. But if 'I remember that p' is not meaningful then it could not be true, and so not incorrigibly true.

At this stage V may propose that we take a crucial step and claim that the verifiability principle may be put to a use which is the inverse of the use to which verificationists are in general supposed to put it. In general, it is assumed that what is to be done is to note that a given sentence is verifiable and hence to classify it as meaningful or else to demonstrate that it is not verifiable and so exclude it from meaningful discourse. But here V suggests that a verificationist may do the opposite: note that a given sentence is undeniably meaningful and hence declare that it must be verifiable. Thus, because we are compelled to judge that 'yesterday it rained' is meaningful, we must also regard it as verifiable. But as it is necessarily the case that no other evidence could be available, we are forced to give our approval to the generally accepted evidence as being sound indicators that yesterday it rained. For consider what kind of evidence is commonly regarded as sufficient to establish the truth of p. I may mention a number of typical examples: (a) I remember myself that it rained yesterday; (b) I can see today that the ground outside is wet everywhere; (c) People tell me that it rained yesterday; (d) I read in the newspaper that it rained yesterday. Clearly the reliability of (a) (b) (c) and (d) as evidence is based on the presupposition that yesterday there was a universe. Thus, if we were permitted to doubt whether the world was more than a moment old we could not possibly confirm that p was true.

Suppose the following objection is raised: admittedly p, which is a sentence part of a meaningful sentence, must be regarded as meaningful. Let it also be conceded that because p is meaningful it must also be confirmable. But perhaps it is confirmable because circumstances are describable in which p would be confirmed as false. For example, after searching as thoroughly as possible for any trace of yesterday's rain and having failed to find any, it could be claimed there is evidence that p is false. One could possibly even go further and claim that if every p-like sentence were confirmed as false, we could well deny the existence of the past.

But V should not find it difficult to meet this objection. After all, the issue raised by the form of scepticism under review is whether circumstances prevailing in the present may be treated as reliable clues concerning conditions prevailing in the past. To the extent that our careful search to find someone who remembered yesterday's rain was unsuccessful, and we found no other clues to indicate it rained yesterday, may be taken as confirming the falsity of p, we can admit that the state of the universe at present is relevant to what happened in the past. In that case there is no good reason to cast doubts on the trustworthiness of those clues that indicate the occurrence of events in the past.

It is clear that V's claim implies that an indefinite number of experiences that may confirm the existence of the past. All those instances that may normally be treated as evidence for the occurrence of some event in the past may be construed as confirming the existence of the past.

5 *Allaying Scepticism Concerning the External World*

It may be pointed out that the method we have adopted to deal with scepticism concerning the existence of the past may also be applied to the more radical scepticism which questions the existence of an external world, such as wondering whether perhaps I merely have impressions that there is something outside me, when in reality there exists nothing apart from my mind, which is the source of these impressions. To allay this form of scepticism V may point out that a considerable number of philosophers have agreed that I may incorrigibly assert that 'I perceive that q' where, for example, 'q' stands for 'there is a table in front of me'. This is so even it is understood that when I am uttering 'I perceive that q', I am not making a direct claim that q actually obtains. That is, my utterance is not taken as a direct assertion that there is something to be perceived only as a report of what is going on in my mind. In other words, I may incorrigibly assert 'I perceive that q', when all I wish to convey is the information that I am undergoing a certain visual or tactile experience. In this case 'perceive' is not used to function like 'know', for in that case I could not be said truly to perceive that which was not the case, but where 'I perceive' functions merely to report a sense impression. The statement then is an avowal of my immediately felt experience, on which I am according to many philosophers an absolute authority. It would be absurd for anyone to attempt to talk me out of it, to try and convince me that I am mistaken, or that it is not a fact that I perceive that q (saying perhaps that it only appears to me that I perceive that q, when of course there is no distinction between 'It appears to me that I perceive that q' and 'I perceive that q').

But if 'I perceive that q' is true, then it may be assumed that 'q' is meaningful. If 'q' is meaningful, it must be verifiable. But what might amount to a verification of q? The following may be regarded as typical examples: (a) I perceive with my eyes that there is a table in front of me; (b) I can touch it with my hands and feel that there is a table in front of me; (c) It is dark and I have not touched anything, but just before the lights went out I saw a table in front of me and not having heard the slightest sound since, I assume that the table could not have been removed; (d) Somebody completely reliable tells me that there is a table in front of me, on the basis of the kind of evidence mentioned in (a), (b), or (c). It is apparent that (a) (b) (c) and (d) are ultimately based on the testimony of my senses as would any other imaginable statements that could be regarded as evidence for q. Thus, if we were permitted to query whether our sense impressions are reliable indicators of conditions in the external world, we could not possibly confirm that q was true.

We have begun this chapter by saying that many philosophers have concluded that the verification principle is invalid because it is undefinable. As I have shown, there is no basis for this conclusion. We have also seen that if we assume the validity of the verification principle, it can perform very important tasks for us: it can allay various forms of scepticism, such as scepticism concerning the existence of an external world and scepticism concerning the existence of the past.

QUESTIONS AND OBJECTIONS

CHAPTER I

(1) Ian Hinkfuss in his *The Existence of Space and Time* (Oxford 1975) claims that there is a fundamental difference between space and time:

> Spatial movement seems to be within our control but our movement through time seems to be inexorable (p. 81).

What is the fallacy in this claim?

(2) In *The Language of Time* (New York 1968) Richard Gale is trying to argue that space and time are dissimilar. On page 215 he says:

> . . . it is meaningful to ask 'Where are you?' to which the answer could be 'I am here': herein it is the direction that the voice comes from that constitutes the answer. However there is no use for the question 'When are you?': 'I am now' has no use in our language. The reason why it is meaningless to ask this question is that it is a precondition for the asking of an A-question that the communicants share the same present, although not the same here, and obviously this precondition cannot itself be the subject of an A-question, as it is when we ask someone 'When are you?'.

What is wrong with Gale's argument?

(3) In the same book, on pp. 214–15, Gale says:

> Every event later than the present will become present and every event earlier than the present did become present, to which the spatial analogue would be that every object in front (to the right, etc.) of me will occupy (become) here and every object in the rear (to the left, etc.) of me has occupied here. But whereas the former is necessarily true the latter is contingent, and what is more is almost certainly false.

Thus he concludes that space and time are dissimilar. Why is his argument incorrect?

(4) In *Problems of Space and Time,* edited by J. J. C. Smart (New York 1964), page 388, Richard Taylor cites the following objection to the thesis that time and space are similar:

> But time is something moving, or flowing, in a fixed direction from future to past and at an unalterable rate; space on the other hand, is everywhere the same and unchanging.
>
> It is for this reason that we speak meaningfully of the passage of time, of the continual recession of things past and the approach of things future. It is because of this too that people naturally think of time as like a great river, engulfing all things in its course. Nothing of the sort, however, is appropriate to the notion of space.

He replies to this that in fact 'there is *no* sense in which time moves'. In explaining this he mentions some of the difficulties we have cited in Chapter Two that attend the notion of a moving NOW. But as we have seen in that chapter, in spite of these difficulties there are philosophers who maintain that the NOW (as distinct from the HERE) moves. How do they deal with Taylor's objection?

(5) In an article 'Space and Time Re-assimilated' (*Mind* 1975) Bernard Mayo examines the question whether space has order and direction in a fashion that parallels order and direction in time. Before we can investigate this problem—Mayo says—it is necessary to ask whether if we take zero dimensional moments as the elements of the one-dimensional time continuum, we are then entitled to say that in parallel fashion the elements of the three-dimensional spatial continuum are two-dimensional surfaces? Mayo thinks we cannot and he says:

> This looks unpromising at first: how can we compare a succession of moments with a series of surfaces? Indeed we cannot, for Euclidean plane surface (p. 578).

Thus, in order to ensure that we have the appropriate counterparts he suggests that we must consider space made up of spherical surfaces, as physical processes such as radiation and wave propagation proceed along a series of concentrically ordered spherical surfaces.

Mayo maintains that if we look on space as a series of concentric spherical surfaces, rather than a series of plane surfaces, we do manage to hit on the right spatial counterparts of the temporal series of moments. The reason is because if we consider surfaces such as a, b, and c, where a is smaller than b and b is smaller than c, then light which is propagated from the common center would reach a temporally prior to when it reaches b and b prior to c.

What is wrong with Mayo's argument?

(6)   On page 14 it is stated that the difference in dimensionality cannot serve as a counterexample to the thesis that space and time are basically alike. But why not? Are we to say that it is merely a contingent fact that time is one-dimensional, but in principle a multi-dimensional time is possible?

(7)   On pages 21–23 it has been claimed that space and time are dissimilar as it is possible in principle to determine that spatially separated events are co-temporal but not that two temporally separated events are co-spatial. It might be objected that surely I can determine, for instance, that I ate breakfast on Sunday and on Monday at exactly the same spot on this planet. Admittedly, the earth partakes in a very complex set of motions and hence I cannot claim to know that the two breakfasts occurred in the same location relative to space itself. But then there are philosophers who would deny that there is any such thing as 'space itself'; according to them, all we have is spatial objects and if they were all removed we would not be left with something called 'space itself', which exists independently of all those objects, but rather would be left with nothing at all. These philosophers might well ask what is it precisely that I claimed to be impossible to determine? Identical position relative to space itself? But that makes no sense, as indeed simultaneity in the context of time itself makes no sense, as 'time itself' does not exist for if there were no events at all no such thing as 'time itself' would remain. But then contrary to what has been claimed in the text, there is no dissimilarity, for in the case of time we cannot talk about the simultaneity of distant events unless we have an event-filled time. But then we can also speak of the co-spatiality of temporally separated events in the context of the earth or any other spatial object.

(8)   Since $speed = \dfrac{spatial\ distance}{temporal\ interval}$, speed is infinite when the

denominator equals zero. It is clear therefore that we get the counterpart of infinite speed by equating the numerator zero; that is, infinitely slow signals are the counterparts of infinitely fast signals. Thus, there seems to be another way in which to eliminate the asymmetry referred to in the previous question. It has been that it is possible to determine that two spatially separated events are co-temporal given that infinitely fast signals (relative to all systems) exist. But then the genuine counterpart of this statement is also true: It is possible to determine that two temporally separated events are co-spatial given that there is a signal with zero speed relative to all systems.

Thus is there no dissimilarity between space and time after all?

(9)   Several people (e. g., R. Swinbourne in his *Space and Time*) have objected to Quinton's argument that the same individual may divide his life and live alternately in two spaces entirely unrelated to one another. They have pointed out that many philosophers hold the view that bodily continuity constitutes the essential criterion for personal identity. Consequently, it is logically impossible for a given individual to 'inhabit' one body and then jump into another body non-contiguous with it even in the same space. It is even more obvious that one individual cannot maintain two bodies in two unrelated spatial systems.

Do we have to agree that on this view of personal identity Quinton's argument fails?

(10)   When we have a statement whose denial is self-contradictory, that statement is said to be necessarily true. In this context the notion of necessity is quite straightforward. However in this chapter the notion of necessity has been used several times in a different sense. For example, 'T' = 'It is possible to determine that two spatially separated events are co-temporal given that infinitely fast signals exist' has been described as necessarily true. Surely the denial of T does not amount to a self-contradiction. Thus it may be asked: in what sense is T necessarily true?

(11)   It has been said that space and time share all those properties that all continua have in common. But it is not entirely certain that either space or time is continuous; people have suggested that discrete, quantized space and time are possible!

(12)   Some philosophers have agreed that it is only with regard to continuum properties that space and time must be similar; nevertheless they maintain the Doctrine of the Similarity of Space and Time. Does such a position make sense?

CHAPTER II

(13)   On page 36 the NOW is defined as '. . . the point in time at which any individual who is temporally extended is alive. . . .' It may well be asked, when we say 'is alive' are we using a tensed 'is'? In that case it means 'is now', and hence the definiens of NOW contains the term NOW and thus the definition is circular. On the other hand if the 'is' in question is tenseless, then all we are saying is that an individual at t_1 is alive at t_1, just as the same individual at t_2 is alive at t_2, and indeed at any t he is alive at that t. Then we are not discriminating in favor of any particular point in time; we treat all instances alike. In this way the NOW loses all its alleged substance.

(14)   On page 38 it is implied that NOWness is a property, yet it would seem that NOW denotes a time, for to the question 'When does E occur'? it may be replied: now. What then is the truth? Is NOW a property of a certain point in time?

(15)   According to the Reichenbach–Smart version of the Russellian position
        E is now = E is simultaneous with this token
It has been objected however that the left-hand side cannot be equivalent to the right-hand side because the former if true expresses a transient truth, whereas the latter expresses a permanent truth. How does one meet this objection?

(16)   It has also been objected that the right-hand side merely expresses the simultaneity of two events, that is, it asserts that the occurrence of E coincides in time with the event of the utterance of a token. It fails however to mention *when* the two events take place. The left-hand side does say when E takes place, namely, now. So the two cannot be equivalent. What is the answer to this objection?

(17) Some philosophers have maintained that the above analysis is involved in a circularity. For example, Paul Fitzgerald in 'Nowness and the Understanding of Time', *Boston Studies in the Philosophy of Science,* Vol. XX (Dordrecht 1974), p. 266, says:

> One problem with the suggested analysis is that it contains the phrase 'this token'. I think that the phrase means 'the token produced here and now' or for short 'the here-and-now token' where it is understood that the boundaries of the here-and-now are so constricted as to admit just one token. But part of the meaning of 'this token' is 'now' the term to be eliminated. So the suggested analysis is not reductive in the sense of eliminating temporal indexical expressions in favor of non-indexicals, 'this token' is itself a temporal indexical.

How would Russellians meet this objection?

(18) Another objection that has sometimes been voiced is that the analysis cannot be correct because the analysan refers to a token that is not uttered when the analysandum is being uttered. In the article cited in the last question on pp. 267–8 Fitzgerald says:

> The analysandum
>
> Event E is occurring now
>
> contains five words. If it is implicitly self-referring, as the suggested analysis implies then it refers to a five-worded token. But the analysans namely
>
> Event E occurs simultaneously with this token
>
> refers to a seven-worded token. Since the token to which the analysans refers is different from the token to which the analysandum refers, and since neither of these tokens is a logical construction the analysans and the analysandum are not logically equivalent.

What is the answer to this?

(19) Another objection to the view that equates the phrase 'at present' with the phrase 'is simultaneous with this utterance' has been that this cannot be so because the proposition 'At present nothing is being uttered' is a contingent proposition, whereas 'At the time which is simultaneous with this utterance nothing is being uttered' is self-contradictory.

(20)   Is it possible that E should have occurred in the past although it is not the case that there was ever a time at which E was occurring in the present?

(21)   We said that statements assigning different A-properties to events may have different truth-values. Can we not say something stronger, namely, that they are contraries?

(22)   J. F. Rosenberg thinks that the definition according to which an A-statement is one which may have different truth-values when asserted on different occasions is wrong. He asks us to consider the following statement:

S = The charging of the Light Brigade is tenselessly simultaneous with the one and only utterance the first ten words of which are 'The charging of the Light Brigade is tenselessly simultaneous with' and the last five words of which are '4:17 P.M. October 3, 1835', and the charging of the Light Brigade is tenselessly simultaneous with 4:17, October 3, 1835.

From the phrase 'the one and only utterance, etc.', it follows that if S is asserted more than once it is obviously false on every occasion of its assertion. Nevertheless, if it is asserted only once and at 4:17 on October 3, 1835, it truly makes the A-assertion equivalent to saying 'The Light Brigade is charging now'. Hence an A-assertion may be made by a statement that cannot have different truth-values when asserted on different occasions.

What is Rosenberg's obvious mistake? Why is there no need whatever to revise the definition?

(23)   On page 46 it is stated that 'being in the future' as such is a property in itself. Why can we not say that, strictly speaking, 'being in the future at t_1' is a property that an event or a moment may

have, and which they may have permanently?

(24) We have discussed a number of objections to the view that there is a moving NOW, objections that are designed to show that this view is incoherent. Some philosophers, in an effort to overcome this difficulty, have claimed that there is such a thing as the moving NOW, although it is not a feature of the physical universe but merely of our mental world. They have compared it to colors that do not exist in the physical universe either, in which there are only light rays with different wave lengths that give rise to different color sensations in our mental world.

Do they succeed in eliminating the difficulty?

(25) Is it not reasonable to hold with Russell that we should adopt it as an axiom that propositions or statements do not ever change in truth-value? If we do adopt such an axiom, do we then have a straightforward argument which proves conclusively that A-statements do not change truth-values, in contradistinction to what I have claimed on behalf of McTaggart?

(26) H. N. Castañeda in his 'Indicators and Quasi-Indicators' in *Amer. Phil. Quarterly* 1967 calls personal and demonstrative pronouns and adverbs such as 'this', 'that', 'I', 'you', 'here', 'there', and 'now' *indicators* when they are used to make strictly demonstrative reference, i. e., when they are used purely referentially, either to single out an item present in the speaker's current experience or to pinpoint a self that is a relation in the cognitive relation evinced by the speaker (p. 85). Reference to an entity by means of an indicator is purely referential, i. e., it is a reference that attributes no property to the entity in question (p. 86).

For example if we compare

(1) I sit on a chair

to

(2) I sit on this chair

we see that (2) attributes no additional properties to the piece of furniture mentioned in (1) except that it indicates on which chair I am sitting.

Are all the words listed by Castañeda 'indicators' in the sense he defined, according to McTaggart? According to Russell?

(27)   Is 'here' the exact spatial counterpart of 'now' or is it not?

(28)   In an oft-quoted article 'Omniscience and Immutability' (*J. of Phil.* 1966) N. Kretzmann claims

> **(K)** A being that always knows what time it is is subject to change.

He explains that to say of any being that it knows something different from what it used to know is to say that it changed. But the being referred to in **(K)** knows that it is now t_1 (and that it is not now t_2) and then it knows that it is now t_2 (and that it is not now t_1). Consequently **(K)** must be true.
    Is Kretzmann right?

(29)   According to McTaggart there are privileged moments. The present moment is privileged in the sense that it is more real than any other moment in the past or in the future. According to Russell the present moment is no different from any other moment—all moments are equal. F. Ferrè claims that Russell's position—which is also A. Grünbaum's position—is untenable:

> . . . suppose . . . that it is true to say of me that as constituting part of my life's career a "now"—awareness event . . . A_1, exists (occurs tenselessly) at clock time t_1 and that it is also true in my life's career that a different "now"—awareness event A_2, exists (occurs tenselessly) at clock time t_2. In the universe depicted by Grünbaum both events have equal claim to existence; neither clock time t_1 nor t_2 has any intrinsic claim to privileged status as more "really" now (whatever, to Grünbaum, that phrase could mean) than the other. Suppose, further, that it is a phenomenological fact that A_2 fills my subjective field of awareness not A_1.
>   Why?
> It will not do for Grünbaum to answer that I am uniquely experiencing A_2 simply because it is clock time t_2. Objectively speaking, on his view, it is (tenselessly) no less clock time t_1, and at time t_1 a "now"—awareness event A_1 no less genuinely exists (occurs) for me. Yet I find myself involuntarily discriminating against t_1 and its associated event A_1 *with no objective ground whatever* to account for this (*British J. for*

Phil. of Science, 1970, p. 279).

Is Ferrè right?

(30)   Richard Taylor in his *Metaphysics* (Englewood Cliffs, N. J. 1963, p. 80) claims that there are certain statements that create difficulties for Russellians:

> Consider, then, our first statement "X is growing older" said of anything whatever, such as a man or a house. One might suppose this to be equivalent to "X occupies an interval of time" or "X exists for more than an instant" or "X endures" or to get rid of the verb, to "X has duration'
>
> These are clearly not equivalent to our original assertion, however, for they might apply to something that is not becoming older at all—to past things, for example, that last more than an instant, such as Diogenes' cup. This cup ceased growing older as soon as it ceased to be, and yet no one could deny that it has existence or duration through an interval of time.

He goes on to explain that any acceptable translation of "X is growing older" must be in terms of what we have called A-statements. Thus we cannot get rid of A-statements.

What does one reply to this on Russell's behalf?

(31)   On page 59 it was pointed out that when all that we are given is that a person is situated at point p_1 and that at both one mile to his left at p_0 and one mile to his right at p_2 dreadful events occur to him at times not revealed to us, then we do not know enough to be able to determine how his attitudes differ toward p_0 and p_2. On the other hand, if we are given that a person's temporal part a, which is at t_1, is situated one day later than t_0 and one day earlier than t_2, and at t_0 and t_2 equally painful events occur, then we know that a experiences fear when contemplating t_2 but not when contemplating t_0. According to McTaggart the basic difference between the temporal situation and its spatial counterpart can be correlated with the fundamental difference between space and time, namely, that the temporal series has a moving NOW. On the other hand, Russell is forced to accept it as a brute psychological fact that we fear certain kinds of future events but not past events.

It might be objected, however, why not account for the difference in attitudes in terms of a difference available to Russell as well, namely, that time has direction and space does not? Thus, t_o and t_2 are asymmetrical with respect to t_1, whereas p_o and p_2 are symmetrical with respect to p_1. t_o and t_2 are asymmetrical with respect to t_1 since t_o is before whereas t_2 is after t_1. On the other hand p_o and p_2 are symmetrical with respect to p_1, for even though p_o is to the left and p_2 is to the right, by turning them around, left and right are interchangeable.

(32)   It might be claimed that in order to account for the difference in our attitude toward the future and the past, Russell need not concede that we are actually moving toward the future; it may be sufficient if he agrees that we have the *impression* of so doing. But then Russell does not deny that we have such an impression. He can also account for the origin of this impression. Suppose a person's temporal part a is at t_1, part b at t_2 . . . part j at t_{10}, and a very unpleasant event is known to a to be going to happen to j. Then a is disturbed, for he has the impression that successive parts of the person whose part a is, are getting closer to t_{10}: b is a step nearer to t_{10}, c is yet nearer . . . i at t_9 is almost there. On the other hand, at t_{11} the temporal part k has a sensation of relief, for k is one step away from t_{10}, and each successive part, $l, m, n,$ etc. is further away from t_{10}.

(33)   We said that Russell would find it impossible to interpret a statement such as 'I wish I was ten years younger.' Some have objected that this difficulty is so only as long as we fail to realize that a person consists of a series of temporal parts. At any time a person expresses a wish, he cannot be considered as a whole, i. e., all his temporal parts taken together cannot be considered to be making that wish, but rather it is a specific part of him, located at the particular position in time at which he utters the wish, that is doing the wishing. We must remember that temporal part a located at t_o is relatively hale and healthy, while b at t_1 is a more worn and deteriorated temporal part. We must realize that it is not a temporally extended whole person who wishes the NOW to be elsewhere, but the somewhat withered b which wishes it was in a more vigorous physiological state similar to the one enjoyed by a located ten years earlier. The wish thus interpreted makes full sense and presupposes no moving NOW.

(34)   I have heard some Russellians try to interpret 'I wish it was t_o' as 'I wish this token were occurring simultaneously with the

events of t_o'. Such interpretation avoids completely any reference to a moving NOW.

Can Russell not be defended in this manner?

CHAPTER III

(35)   McTaggart has shown that there is an asymmetry between A-statements and B-statements, in that the latter may be defined in terms of the former but not the other way round. May not one point at another asymmetry as well: the A-statement 'E_1 and E_2 occur now' entail the B-statement 'E_1 and E occur simultaneously' but not vice versa?

(36)   On page 75 two conditions are given which are said to be required in order that a statement should qualify as an A-statement. But in fact there is only one condition, for any statement satisfying (i) necessarily satisfies (ii) as well! Let us suppose p satisfies (i), i.e., as it stands it may undergo changes in truth-value; then it logically follows that there is at least one way in which it may be broken up into components such that all of them are A-statements i.e., (p v q) & (p v ~ q) where q is an A-statement!

(37)   Stephen E. Braude in his article 'Tensed Sentences and Free Repeatibility', *Philosophical Review* 1973, discusses at length the various attempts that have been made in the past to define a tensed sentence (or statement). He rejects the attempt to identify tensed statements with A-statements. First of all he says that A-statements refer to events, whereas tensed statements need not. Secondly, he argues that 'John will leave home after Mary returns with the car' unmistakably expresses a B-relation between two events, yet it is tensed. Does the attempt to identify tensed statements with A-statements indeed fail?

(38)   It has been claimed that if E is an unextended event, it cannot accommodate the incompatible predicates P and not P. But suppose at t_1 I remember E but by t_2 I forget it and let 'P' stand for 'remembered by me'. Then at t_1 it is true that E is P and at t_2 it is true that E is not P, is it not?

(39)   On page 92 it is argued that the two statements

and

'P is Q' is true when asserted at M $_1$ . . . . . . (d)

'P is Q' is not true when asserted at M$_2$ . . . . . . (e)

cannot be both true, because if P is a temporally unextended particular it either has Q, in which case e is false, or it does not have Q, in which case d is false.

But why not reply to this, that whereas indeed *d* and *e* could not both be true as long as Q stood for any ordinary property, in the exceptional case of A-determinations, which are very special properties, they may both be true?

(40)   What is said on page 101 suggests that the definition of a benign regress is that it is a regress in which we benefit by learning something new after each step we take. It is, however, easy to offer a counter-example: everyone would agree that the regress generated by the truth that for every true proposition it is true that it is true, is a benign regress. Yet nothing new is learned as we progress along the regress.

(41)   How by making use of a certain idea of Broad mentioned in Chapter Two could McTaggart's difficulty be overcome?

CHAPTER IV

(42)   Most philosophers think that arguments designed to establish fatalism are fallacious. Consider the following argument, reported to have been devised by British philosophers during World War II at the height of the air raids, which concluded that there was no point in going to the air shelter. The argument was based on the following three premises:

(1)   If I am going to be hurt in this air raid then I am going to be hurt no matter what precautions I take.

It is impossible to deny that (1) is necessarily true, for if it is given that it is true that I am going to be hurt in this air raid then this proposition remains true no matter what other proposition is also true. Now (1) is translated into

(1')   If I am going to be hurt in this air raid, then all precautions are useless.

For the same reason that (1) is necessarily true, (2) is also necessarily true where

> (2)   If I am not going to be hurt in this air raid, then I am not going to be hurt in this air raid no matter what precautions I neglect.

(2) may be translated into

> (2′)   If I am not going to be hurt in this air raid, then all precautions are unnecessary.

By the Law of Excluded Middle we also have

> (3)   Either I am going to be hurt in this air raid or I am not going to be hurt in this air raid.

Clearly (1′), (2′), and (3) logically imply:

> (4)   Either all precautions are useless or all precautions are unnecessary.

It is, of course, pointless to take precautions in either case. It would be useless to argue that it is not pointless to take precautions, for statistics show that the proportion of injuries was much higher among those who did not take precautions during air raids than among those who did. No matter how well established the statistical results, as long as we have a logical proof that there is no point in taking precautions the difficulty remains. The difficulty can only be solved by either showing that the conclusion does not follow from the premises or that one or more than one of the premises is not necessarily true.

What is the solution?

(43)   Consider the following brief presentation of an argument devised by R. Taylor in his *Metaphysics,* Chapter 5, to prove that fatalism is true. Taylor assumes that we are all fatalistic with respect to past events. And we shall not deny him this assumption. Experience does show that the correct attitude toward the past is a fatalistic one, because in practice nobody claims that we have now the power to affect the past; the past is the way it is independently of anything we may now do. Taylor begins by presenting a formal argument to show why the fatalistic attitude to past events is required.

Let P = A naval battle occurred yesterday.

    P′ = A naval battle did not occur yesterday.

    S = I perform an act of seeing a newspaper headline that P

    S′ = I perform an act of seeing a newspaper headline that P′

We shall assume that P is necessary for S and P′ for S′. Then the following three premises are true.

(1) If P is true it is not within my power to do S′ (for in case P is true, a condition was missing which was essential for my doing S′, namely, of there being no naval battle).

(2) If P′ is true, it is not within my power to do S [(2) is true for a reason similar to why (1) is true].

(3) Either P is true or P′ is true.

From which it follows that:

(4) Either it is not within my power to do S or it is not within my power to do S′.

This amounts to an expression of fatalism with respect to the past. Taylor then goes on to argue that an exactly parallel argument exists to establish fatalism with respect to future events. Let

Q = A naval battle will occur tomorrow.
Q′ = A naval battle will not occur tomorrow.
O = I give an order that Q.
O′ = I give an order that Q′.

Let us suppose that I am a naval commander whose orders are absolutely certain to be carried out and, therefore, if O is true then Q is true and if O′ is true then Q′ is true. We again have

(1′) If Q is true, it is not within my power to do O′ (for in case Q is true, a condition is missing which is essential for my doing O′, namely, of there being no naval battle).

(2′) If Q′ is true, it is not within my power to do O
[(2′) is true for a reason similar to why (1′) is true].

(3′) Either Q is true or Q′ is true.

(4′) Either it is not within my power to do O′ or it is
not within my power to do O.

Thus, fatalism with respect to the future is established in exactly the
same manner as we established fatalism with respect to the past.
   Is there really a parallel between the two arguments?

(44) It might be objected that it makes no sense to stipulate that
the rotation of X at t_2 is the cause of X becoming magnetized at t_1.
It might be asked: when at t_1 we were watching Y emit radiation,
was X magnetized or not? If X was not magnetized at t_1, then we
cannot say later that X was magnetized at t_1 because of its rotation
at t_2; if X was magnetized at t_1, then the causal effect of the rotation at t_2 is not needed, is it?!

(45) Does the law of exportation apply to causes? Is it true (as
implied on page 121) that if A and B together cause C, then A
causes B to cause C?

(46) It has been claimed that one of the difficulties Interpretation F had to contend with was that a physical event such as a, i. e.,
that X becomes magnetized at t_1, of which Tom may not be aware
at all, may be the cause of a mental event such as Tom deciding at t_2
to rotate X. But the same kind of difficulty seems to face those who
hold Interpretation B. It is given that without a it would not be the
case that α; that is, without X being magnetized at t_1, it could not
be true that Tom decides to rotate X at t_2. In other words Tom cannot decide to rotate X at t_2 unless X has been magnetized at t_1. So
the question that faces those who hold Interpretation B is: why is it
that a mental event such as Tom's decision to rotate X requires the
occurrence of a kind of event such as X being magnetized at t_1?

(47) It has been shown that the fact that an object X rotating in
an electric field (α & β) is the cause of X's earlier magnetization (a)
can emerge increasingly more clearly as we made more and more
relevant observations. What has not been given, however, is a

description of the feature which distinguishes a cause from its effect. Surely there must be some distinct feature by virtue of which one of the pair of events causally connected is regarded as the cause and the other the effect, a feature which is ultimately responsible for leading us to the conclusion that α & β are the cause of a and not vice versa. What is it?

(48)   Martin Gardner in the Mathematical Games section in the *Scientific American* discusses a case of backward causation. He asks us to suppose that there are two transparent boxes in front of me and I have two choices: (a) take box II only; (b) take both boxes. Box I contains one thousand dollars. In box II is a sheet of paper with a number written on it of so many figures that I am unable to determine on sight whether it is prime or composite. There is a person, P, who will give me one million dollars if the number is prime. P is an infallible predictor, and it is he who has written the number on the paper in the box according to the following principle: if P predicts that I will take box II only, he writes down a prime number; if he predicts that I will take both boxes he writes down a composite number. Although we may assume that if I take box II only the number will turn out to be prime and if I take both boxes it will turn out to be composite, it would be wrong, says Gardner, repeating the views of William Newcomb, to construe my choice as the cause which actually determines the nature of the number on the paper. He gives the following explanation:

> "Obviously you cannot by an act of will make the large number change from prime to composite and vice versa. The nature of the number is fixed for eternity." (*Scientific American,* July, 1973, p. 108.)

What is the fallacy with his reasoning?

(49)   We have confined our attention to a predictor whose competence has been established through his vast record of successful predictions in the past. Consequently, on Interpretation 1 when we arrive at a contradiction we are able to conclude that in spite of the overwhelming inductive evidence supporting the claim that the predictor is virtually perfect, we now have deductive proof nullifying the weaker, inductive proof, that in fact he has no competence at all. It may however be asked: why not start out differently and postulate that the predictor is absolutely perfect as, for example, if it is given that he is an Omniscient Being?

Along similar lines we may ask: if on Interpretation 2 we stipulated that the correlation between T_i and C_i was not merely exceedingly high but absolute, then surely we could no longer maintain that under all circumstances CH should try to do C_2, because if P_1 then it is certain to be T_1, which in turn definitely ensures that C_1 thus doing C_2 is simply out of the question?

(50) It was claimed that the opinion of the Perfect Judge must be correct. But we should remember that his view that it is best to do C_2 may be held in two different contexts: in the context of an empty box, in which case he holds it is better to do C_2, because in that case the player gets at least one thousand dollars; or in the context of a full box, in which case he holds that it is better to do C_2 than C_1 and gain \$1,001,000 rather than just \$1,000,000. But if we assume that the predictor is perfect, then by doing C_2 the player is sure to bring about that he is better off in the context of an empty box, whereas if he does C_1 then he brings about that he is worse off in the context of a full box. Now it so happens that it is ultimately preferable to be worse off in the context of a full box, where the player fails to get the \$1000 of box I but gains the \$M that is in box II, than to be better off in the context of an empty box, by which the player ends up with no more than \$1000. Thus, ultimately it is preferable not to act according to the opinion of the so-called Perfect Judge, is it not?

(51) In order to avoid the contradiction that on the one hand inductive evidence shows that it is best to do C_1, whereas on the other hand the opinion of the Perfect Judge is that it is best to do C_2, we might say that when the predictor who has been described as a Perfect Judge believes that he has put \$M in the opaque box and expresses the opinion that it is best to do C_2, and the player accepts his opinion and acutally does C_2, then it will turn out that there was no money in the opaque box to begin with, and the predictor misremembered it when he thought he had put money in box II. Thus, his opinion is unreliable.
Is there anything wrong with this argument?

(52) It was stated that even a slightly competent predictor of human choices does not exist. But surely this is contrary to common experience. All of us are fairly competent at predicting the free choices of people with whom we are well acquainted. I am not endowed with any special abilities for prediction; nevertheless, if I have known a person for many years I am quite likely to foretell correctly how he will react in certain situations which are similar to those in which I have seen him many times before. Of course he may surprise me and at some time act differently from the expected, but the probability is high that this will not happen. Hence, I am not an entirely incompetent predictor?

(53)   Max Black in his article "Why Cannot an Effect Precede Its Cause" in *Analysis,* 1955 discusses the case of a man who can invariably predict (A) the final outcome of the tossing (T) of a penny. He considers the question whether it is possible to maintain that T is the cause of A, and hence that we have a case of backward causation here. He rejects this possibility by saying that T cannot be the cause of A, because on any occasion when A occurs one could simply prevent T from occurring, showing that A does not require T.
What is wrong with this argument?

CHAPTER V

(54)   Norman Malcolm, and other philosophers influenced by him, have argued that we have incontrovertible evidence that there was a past. Their argument is briefly as follows: If we encountered a person whose every statement about the past was false, we would justifiably come to the conclusion that that person did not understand the notion of 'past'. By counterposition it follows that if a given person understands the notion of 'past', it is not the case that every statement he makes about the past is false. But we regard ourselves as people who understand the notion of 'past'; hence, at least some of our statements about the past must be true. But if there was no past, none of our statements about the past could be true. Hence, there was a past.
What is wrong with this argument?

(55)   Some have argued that there must have been a past for the following reason: I am able to describe a great deal of what goes on in the world. For example, I can state that there is a table in front of me. But how would I know that what was in front of me was a table unless I acquired through experience the knowledge that this is so? Thus, there must have been time in the past during which to acquire the requisite knowledge through experience.

Is this a good argument?

(56)   We have referred to the oft-repeated objection to verificationism that the criterion of verifiability is undefinable. There
have, however, been other objections as well. One well known
criticism of the concept of verifiability is that it is closely tied to the
notion of observation, as a sentence is factually meaningful if it is
either expressing an observation-statement or there are
observation-statements which, if true, would confirm or disconfirm it, where 'confirm' and 'disconfirm' are expected eventually to
be fully explicated. Now whatever the ultimate explication of the
term 'confirmation' may be, it is obvious that the factual meaningfulness of sentences is essentially tied to observation-statements.
It is supposed then that we recognize an observation-statement
when we see one and know how to distinguish it from what is, in
contrast, a theoretical statement. But, in view of the theory-
ladenness of all discourse and various other facts, this is by no
means the case. Some may claim that although we cannot actually
observe that A receives more radiation from the sun than B, we can
infer this from the observation that A reaches a higher temperature
when placed in the sun than B. Others, however, will deny that we
can observe even this much and will insist that all we can observe is
that, in a given instrument which we call a thermometer, the mer-
cury column reaches a higher level when in contact with A than
when in contact with B, the rest is theoretical inference. Yet others
may go even further and disallow 'mercury-column' to pose as an
observation-term. All that we may observe is that a silvery line is
longer in one case than in another.
      The claim that verificationism cannot get off the ground because
of the essential uncertainty attaching to the term 'observation-
statement' has been made a number of times. A recent version of it
is by Swinbourne. His attack is relatively moderate, as he refrains
from claiming that the notion of 'observation-statement' is so hazy
as to be vacuous. He merely states that the notion is ambiguous
enough to not be helpful in distinguishing between all that is factually meaningful and what is devoid of such meaning. He says:

> So although men may be agreed *by and then* about which
> statements are observation statements, I see no reason to
> suppose that the degree of consensus is vastly greater here
> than over which statements are factually meaningful. And if
> that is so, the confirmationist principle is not going to be of
> great help in clearing up the latter ('Confirmation and Fac-
> tual Meaningfulness', *Analysis* 1973, p. 73).

What is the answer to this objection?

(57) In the article just quoted, Swinbourne makes another interesting criticism calculated to demolish verificationism. He claims that it may simply be demonstrated that there is an indefinite number of propositions which share a certain form and are so constructed that they cannot be verified under any circumstances and yet they are undeniably meaningful. One such example is:

> p_1: Among possible claims about the prehuman past which the best evidence ever to be obtained by man makes highly improbable, some are nevertheless true.

Everyone will be impelled to agree that p_1 is meaningful. For let q be a particular highly improbable claim about the prehuman past. Then if we properly understand what 'highly improbable' means, we know that it is different from 'false'. It therefore makes sense to say that q is highly improbable but not false. But if that makes sense, then surely p_1, which claims merely that there is at least one proposition which is like q, must also make sense.

Now to the claim that p_1 is in principle unconfirmable. What would amount to the confirmation that p_1 was true? That p_1 was true would be supported if we confirmed Q, where Q asserts that q (which is a claim about the prehuman past), although highly improbable, was nevertheless true. This, however, seems impossible. Q, after all, is a conjunction of two statements *a* and *b* where:

> a: q is highly improbable by the best evidence ever to be obtained by man.

> b: q is true.

Q is true if both its conjuncts are true. If, however, we should confirm *a* to be true, that would amount to the disconfirmation that *b* was true and vice versa. It is just not possible that there should be evidence that Q was true and therefore no evidence that p_1 was true. On the other hand, that p_1 cannot be disconfirmed follows simply from the fact that it is an existential statement and such statements in general cannot be falsified.

What is wrong with this objection?

(58) There are many people who are impatient with verificationism and believe that it is quite easy to show its inadequacy. I have heard it said that confirmability in principle is not a plausible criterion of meaningfulness. After all, 'It is raining but no one has any reason to believe that it is' is not confirmable in principle, but it is certainly meaningful.

Do we have here a refutation of verificationism?

(59)   There have been some philosophers, who unable to reconcile themselves to the fact that such a simple solution should exist to a problem that kept confounding philosophers for so long and that I have indeed succeeded in providing a straightforward definition of 'confirmable in principle' have objected that my definiens contain the term 'coherently describable' which itself calls of an explanation. One of them asked: can we for instance coherently describe someone coming up within formalized arithmetic of a proof of Goldbach's conjecture? Our hesitation as to how to answer this question correctly shows that the term 'coherently describable' is problematic.

Have they been successful in identifying a real difficulty with my definition?

ANSWERS

CHAPTER I

(1) Hinkfuss wishes to point out that spatial movement seems to be under our control, by which he means that we can decide whether to cover a small region or a large region of space during a given interval of time. Spatial movement is to be understood in terms of space covered during a certain amount of time (and not, of course, during a certain amount of space!) Obviously, then, for temporal movement truly to be a counterpart of this, we have to speak of time being covered during a given amount of space. There is nothing inexorable about this. I can decide, no less than in the case of space, whether I wish a small stretch of time or a large stretch of time to be covered at a given spatial interval.

(2) First of all when we ask someone, 'Where are you?' we mean to ask, 'Where are you at time t?' and often t stands for the present time. We may perfectly well ask, 'When are you at place p?' which is the counterpart of the first question, and by p we may denote the place we are at.

Secondly, communicants need not share the same present. I can send today a letter to someone asking 'Where are you?' He may receive the letter a week later, and I may intend to ask him where he is at the time of receiving the letter.

(3) It is clear that the correct spatial counterpart to the temporal statement would be something such as 'Every event to the front of here is here at a place in front of here, and every event to the rear of here is here at a place to the rear of here'. This statement is also necessarily true.

(4) First of all it should be pointed out that the present objection is unlike the objection raised by Hinkfuss, for we are comparing genuine counterparts: the NOW is moving in time, the counterpart of which would be the HERE moving in space, which is not the case. Hence, we are dealing here with a genuine dissimilarity.

The answer, however, is that it does not matter at all, for if the NOW is moving then this is a characteristic temporal property of time; that is, time does not have this property simply by virtue of it being a continuum. As it is not a continuum property we are dealing with here, time may very well have it, while space lacks it. It was really not necessary for Taylor to convince his readers that the NOW is not moving, for even if it does there is no difficulty.

(5)   Mayo errs by thinking that we cannot compare a succession of moments with a series of Euclidean plane surfaces. The truth is that both are full-fledged successions, possessing all the properties that are possessed by ordered series of elements succeeding one another. That this is undoubtedly so is evidenced by the fact that if x, y, and z are elements of either the temporal series of moments or the spatial series of plane surfaces and '$x < y$' denotes 'x precedes y' or 'y succeeds x' then it is jointly true that:

(1) $\sim (x < y) \leftrightarrow [(x = y) \vee (y < x)]$
(2) $x \neq y \leftrightarrow [(x < y) \vee (y < x)]$
(3) $[(x < y) \,\&\, < z)] \rightarrow x < z$

What then is Mayo worried about? Surprisingly enough, he is worried that the plane surfaces do not have to each other, in addition to the relationship of succession in a general sense appertaining to all ordered series of elements, the relationship of 'before' and 'after' in a specific temporal sense!

But the crucial point to realize is that in the case of the spatial elements p_1 and p_2, for example, if p_1 is before p_2, then the sense in which this is so, apart from the general one of precedence that applies to elements in all ordered series as defined by (1), (2) and (3), must be in a purely spatial sense. The characteristically spatial notion of precedence is just as exclusively spatial in nature as the temporal notion of precedence is exclusively temporal. In the case of the temporal elements t_1 and t_2, if t_1 is before t_2, then the specific temporal relationship of t_1 to t_2 is, of course, in no way dependent on any characteristic features of space or, for that matter, of some other particular series. Indeed, the specifically temporal notion of precedence is, according to some, a primitive notion which is undefinable but may be conveyed to a person only through direct demonstration, as when a flash of lightning precedes thunder by remarking 'lightning before thunder.' According to others (e. g., McTaggart, as we have seen in Chapter Three), it may be defined, but only in terms of other purely temporal predicates such as past, present, and future, i. e., t_1 is before t_2 if it is ever true that t_1 is in the present, while t_2 is in the future.

It should be clear that in every sequence there is precedence in the general sense as well as in a specific sense unique to that sequence and not to be borrowed from some other sequence. As is known, the succession of integers constitutes an ordered series where, for example, the number 8 precedes the number 11. But it would amount to a complete misunderstanding to say that the sense specific to this series in which 8 precedes 11 is that 8 is reached *earlier in time* than 11 when integers are counted. No, the relationship is a pure mathematical relationship, which may be explicated in terms of other mathematical relationships, i. e., by saying that 8 precedes 11 because 8 is a smaller number than 11.

Similarly, then, the specific spatial sense in which plane p_1 may be said to precede plane p_2 has to be explicated in purely spatial terms. For example, I may face a certain direction and say that for any two planes p_1 and p_2, 'p_1 is before p_2' = 'p_1 is behind me and p_2 is here or ahead of me, *or* p_1 is here and p_2 is ahead of me *or* p_1 is further behind me than p_2 *or* p_2 is further ahead of me than p_1'. Another way in which to help a person realize that p_1 is spatially before p_2 is by directing his attention to various pairs of objects and pointing out to him which one precedes the other. But obviously spatial precedence does not have any temporal connotations, just as in no sense is temporal precedence a spatial one.

(6)   Whether in principle a multi-dimensional time is possible is a complex and disputed issue. For example, Judith Thomson in 'Time, Space and Objects', *Mind*, 1965 suggested conditions under which a two-dimensional time could materialize. However, R. E. Nusenoff in his 'Two Dimensional Time', *Philosophical Studies*, 1976 disputes her claim. In order to avoid this whole issue I have suggested that we consider the question whether one-dimensional space is or is not similar to time.

(7)   We may very well assume that neither time itself or space itself exists, and yet there is a basic asymmetry. Even in a non-empty space we cannot talk about co-spatiality; we are compelled to designate a given system which we regard as stationary. Two temporally separated events that are co-spatial with respect to the earth are not co-spatial with respect to the sun, and vice versa. Thus, we must arbitrarily specify a given point as the origin of our spatial co-ordinates. On the other hand once we have events we do not need to designate any event as the origin or our temporal co-ordinates. Two spatially separated events are either co-temporal as such or

they are not. It cannot be the case that they are co-temporal relative to some specific event but not relative to others.

(8)   Admittedly, co-spatiality is determinable under the circumstances described. But we must remember that a universe in which there was a signal that travelled at zero velocity relative to all systems would be a tremendously restricted universe: it would have to be a universe in which everything was at complete standstill. No parallel restrictions, indeed no restrictions whatever, are required to be placed on a universe in which the co-temporality of distant events are determinable. This in itself amounts to a dissimilarity between space and time.

(9)   The essential point in Quinton's argument is not affected if we subscribe to the view that bodily continuity is a sufficient and necessary condition for personal identity. Although he would be forced to interpret differently his story concerning the person who seems to have gone to bed in England and subsequently found himself awaking on a tropical island located in a different spatial system, he could continue to maintain that it is in principle possible to secure evidence that two unrelated spatial systems exist. Instead of one person, he will have to speak of two persons A and B. When A has memories of having gone to sleep in England and subsequently of having awakened on a tropical island situated in a different spatial system, we shall interpret this as A having veridical memories of experiences to which not he, but B dwelling in England, was subject. Similarly, B has memories of experiences undergone by A, including the experience of A's remembering B's experiences, which B is now in a position to verify.

(10)   The necessity employed in this chapter is conceptual necessity. Given our concept of time, we understand without having to resort to any experiments that T is true. In order to agree to this it is not required that we disregard D. A. T. Gasking's warning, in the context of his discussion of Quine's criticism of the notion of analicity, that we must not forget that our concepts continually undergo changes.[1] Thus, in, say, one hundred years, with the great progress of science, it is possible that our notion of time will undergo such fundamental changes that T may not appear to be

1. "The Analytic-Synthetic Controversy", *Australasian Journal of Philosophy*, 1972.

true. However, in the way in which we now understand the concept of time, it appears clear that T must be true.

(11) It is not vital for us to assume that space and time are continua. My point is that they must share certain mathematical properties. Should one of them have the order type of the rationals and the other of the reals, they still would have to have in common all the properties these series have in common.

(12) Such a position is not self-contradictory. It could be maintained with some plausibility that a proposition asserting that time has a given property may be necessarily true if it refers to a continuum property but not otherwise. The reason is that continuum properties belong to all continua with mathematical certainty. But a proposition assigning some noncontinuum property to time cannot be necessarily true, for it is always an empirical question whether time has this or that (noncontinuum) property. Consequently, the Doctrine of the Similarity of Space and Time must be true without exception, for any necessarily true or necessarily false proposition concerning time must be referring to a continuum property, in which case the spatial counterpart must be true or false, respectively.

Those who would disagree would claim that there are propositions assigning noncontinuum properties to time which are also necessarily true or necessarily false. These philosophers hold that not all necessities are logical necessities; some are conceptual necessities. For example, the proposition 'If E_1 and E_2 occur at the same time at different places then it can be ascertained that they occur at the same time provided there are infinitely fast signals' would be regarded by such philosophers as necessarily true. The truth of the proposition does not follow, of course, from any of the axioms of logic of mathematics. But by understanding the meaning of the terms employed and thinking about the matter, one is able to determine that this is a true proposition.

CHAPTER II

(13) When we translate 't is now' as 't is real and alive', 'the 'is' we employ is a tensed one. But in order to explain what that means we do not say that 't is real and alive now' but rather that the statement 't is real and alive' when asserted at t is true and when asserted at any other time it is false.

(14)   There is no contradiction, for when we say 'E occurs now' we may mean that E has the property of being alive as well as that E occurs at the point in time which is alive and real.

A useful imagery is provided by Gale who suggests that we think of the NOW as a moving spotlight which illuminates successively different moments along the series of time. Thus the NOW may be referred to as: (1) a particular (as on page 51) meaning the moving spotlight; (2) a property of a moment, or of an event, or of a temporal part of an extended process or object, namely, the property of being illuminated; (3) a point in time, namely, the moment which is being illuminated.

(15)   The very point of the Russellian analysis, of course, is to assert that there are no such things as transient truths. The right-hand side means nothing more or less than the left-hand side. The left-hand side, if true, expresses a permanent truth.

(16)   As we have already said, according to Reichenbach and Smart 'E is now' means nothing more or less than 'E is simultaneous with this token'. To the question 'when does E occur?' the answer provided by the left-hand side is not 'now', for 'now' does not stand for a given moment. The answer given is: at the time when this token is being uttered. The same answer is being provided by the right-hand side.

(17)   It is to be maintained that there is no circularity because my saying 'this token' is not to be translated into 'the here-and-now token.' It is not to be translated into any phrase at all. By saying 'this token' I am merely executing an act of pointing. The listener is at once bound to understand that I cannot be pointing at anything else but at the very token being uttered at the moment. Thus, the token is not described to the listener by any predicate at all; it is pointed out to him.

(18)   Fitzgerald misses the point of the analysis, which is to instruct us that the term 'now' is to be treated as an abbreviation and is itself to be taken to stand for 'simultaneously with this token'. Thus 'this token' refers to the original five-worded token.

(19)   Reichenbach and Smart do seem to be committed to the view that 'At present nothing is being uttered' is also self contradictory. If this strikes us as strange, it is only because we have been tacitly assuming a different analysis of 'at present'.

(20)   One might make out a case that Reichenbach and Smart are committed to the view that such a thing is possible. It could be claimed that if E occurred in the distant past when there were no sentient beings and no utterances, then E occurred at some time t where t was in the past (i.e., t occurred before this utterance), yet E had never been in the present, because it had never been the case that E was simultaneous with any utterance.

(21)   It depends. If the events in question are unextended, unique events, then they must be contraries. Thus, 'the first man landing on the moon was American' is necessarily a contrary of 'The first man landing on the moon will be an American'. This is not the case with extended events, for 'the SALT negotiations have been going on in the past' and 'the SALT negotiations will be going on in the future' are not contraries. Neither is it so in the case of non-unique events. 'There was a flash of light' is not a contrary of 'There will be a flash of light'.

(22)   The definition may remain as it stands for there is no problem here. S may not be an A-statement yet it entails an A-statement. There is nothing extraordinary about this. For example, 'It is always ϕ-ing' which is certainly not an A-statement entails 'It is ϕ-ing now'.

(23)   The property of 'being in the future at t_1' is exactly the same as 'being later than t_1'. When it is said of an event E that it is in the future at t_1, what is being asserted is the existence of relationship between E and t_1, and this relationship is a permanent one. But when it is said that E is in the future, as such, then we are not talking about a relationship between E and a fixed time but about the relationship between E and the moving NOW; we are asserting that E is later than the NOW. This relationship is subject to change.

(24)   No. It is of no help to declare the moving NOW to be no more than a feature of our mental world. Mental facts, just as

physical facts, must be coherent. If indeed the Russellians are right
that it is impossible to maintain without inconsistency that there is
a moving NOW, then it is also not possible to maintain that there is
a moving NOW in our mental world.

(25)   Whether or not it is possible for the same statement to
have different truth-values when made on different occasions is not
something to be stipulated but rather something we have to find
out. Even Russell should not be assumed to have adopted it as an
axiom that statements do not ever change in truth-value but rather
that he simply could not conceive of an example in which they did.
All of us, regardless of what position we hold with respect to the
question of temporal becoming, have to look hard and see whether
it is ever possible to say exactly the same thing on different occasions, and yet on one occasion we would be speaking the truth and
on the other one we would not. If we find this to be possible, then
we are forced to admit that the very same statement may undergo
changes in truth-value.

According to all those who deny the existence of A-statements,
such a possibility is not conceivable. According to McTaggart,
however, 'E is now' translated into 'E is at the point in time at
which any individual who is temporally extended is alive'
characterizes something uniquely as possessing the property of
'having extended particulars coming alive at it'. Consequently, 'E
is now' provides exactly the same characterization of what it refers
to when uttered at different times. We are compelled, therefore, to
concede that statements change truth-values.

(26)   It seems clear that according to McTaggart the word 'now'
is not to be counted among indicators. For compare

              (3) I sit on a chair some time.
       with
              (4) I sit on a chair now.

It is clear that the difference between (4) and (3) is not merely that
(4) indicates the specific time at which I sit on a chair while (3) does
not. (4) identifies the time in question by means of a definite
description, that is, by describing it as '*the* point in time at which
any individual who is temporally extended is alive and real'. In
other words the reference to time in (4) is not purely referential, for
it does attribute a property to the time in question, namely, that of
having at it everything come alive.

But even according to Russell it would seem that 'now' must not be counted among indicators. On the Reichenbach–Smart version where

E is now = E is simultaneous with this utterance

(4) identifies the time it refers to as having the property of being simultaneous with a specific utterance. Similarly, according to the well-known Grünbaum line, which assigns simultaneity to E with a certain mental event rather than a given utterance, 'now' cannot be regarded an indicator in the required sense.

It does not follow however that Castañeda is mistaken; in fact, he represents a novel view, one worthy of attention. He may be taken as holding that 'E occurs now' is to be interpreted as 'E occurs at this time', the speaker pointing at the time in question by means of performing the utterance at the time pointed out. 'E occurs at this time' does *imply* that E occurs simultaneously with the utterance of it, because the speaker does pinpoint the time of the utterance by means of his utterance made at that time. The speaker however does not *assert* this. Hence (4) does not assert the attribution of any additional properties to the time of my sitting on a chair; it merely identifies the time in question referentially. On Castañeda's view 'E occurs now' when uttered at different times makes different statements, because the 'this' points at different times. Also, he would be committed to the view that there are no changes that may be called genuine changes in the sense required by McTaggart. Ultimately then, Reichenbach, Smart, and Grünbaum as well as Castañeda are in agreement on the crucial question of A-statements: There are no such statements according to any one of them.

Returning to the question of indicators: if 'E is here' is translated as 'E is co-spatial with me', then obviously 'here' is not to be regarded as an indicator either. Naturally, Castañeda would deny this as being a correct translation. Once more, 'E is here' translates according to him into 'E is at this place', where the demonstrative 'this' points at the place where I am uttering these words.

(27)   (i)   According to the Reichenbach-Smart interpretation it is clear that it is. For just as

E is now = E is simultaneous with this utterance

in exactly parallel fashion we may say that

E is here = E is co-spatial with this utterance

(ii)   According to Grünbaum's interpretation matters are not entirely clear. When I say that E occurs now, Grünbaum takes me to assert that E is simultaneous with a specific complex event of mine which he carefully describes. Would he say something parallel about my assertion that E occurs here? It does not seem plausible that 'here' is mind-dependent in this sense. After all a sign could be placed at the North Pole saying 'The North Pole is here' and the sign would be truly indicating where the North Pole is. Thus, the sign could be telling the truth even though there was no minded being within hundreds of miles, and thus there were no mental events of any sort with which the sign was co-spatial.

But then of course we may also ask: is it clear why 'now' had to be declared mind-dependent? A world is logically possible in which there are no sentient beings at all but there is a clock which announces every hour 'Now it is X years Y days and Z hours since the time the universe came into existence' (where years and days are determined by the movements of the planet on which the clock is situated.) There would seem no objection to claiming that each announcement of the clock corresponded to the truth, even though there were no mental event required by Grünbaum to be co-temporal with these announcements anywhere in the whole universe.

(iii)   It is obvious that according to Castañeda, in the same way in which 'now' indicates the time of my utterance and to which my utterance points, 'here' indicates its place.

(iv)   According to the view that 'E is now' is an A-statement in the sense defined before, 'now' and 'here' are not counterparts. 'E is here' can be taken as 'E is where I am' or 'E is co-spatial with this utterance.'

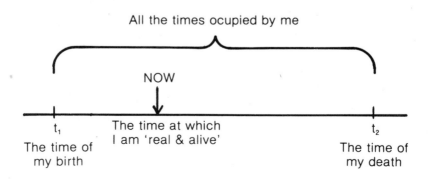

On the other hand, 'E is now' cannot be taken to mean 'E is when I am' simpliciter. If my lifetime stretches from t_1 to t_2 then I am occupying all the points in between. The unique point that is at present is a privileged point at which I am 'alive and real'. Thus, whereas the phrase 'when I am' picks up an interval of many years, it is only the term 'now' which refers to a single point in time.

(28) Not according to Reichenbach, Smart, Grünbaum, or Castañeda. According to the first two authors, for example, if X knows that it is now t_1 X knows that his utterance (at t_1) is simultaneous with t_1. His knowledge of this fact will never need revision; it is simply not true that the particular utterance just referred to will ever be simultaneous with t_2. He may also know at t_2 that it is correct to utter then 'it is now t_2', meaning that the utterance (made at t_2) is simultaneous with t_2. This again is a piece of knowledge that will not ever require modification.

According to Castañeda when X utters at t_1 'It is now t_1', he is pointing at the current time and designates it correctly as t_1. It will never turn out that he has done at t_1 anything inappropriate. At t_2 he is pointing at a different time and correctly identifies it as t_2 and by doing so he is not retracting anything he said at t_1. Thus X's knowledge undergoes no change whatever and much less is X himself subject to any transformations.

Only according to the position initially held (and subsequently abandoned when arriving at the conclusion that time is unreal) by McTaggart might there be any room to entertain (K).[2]

(29) Ferrè has not discovered a difficulty in the Russellian position. Indeed, t_2 is no more privileged per se than t_1. But *at* t_2, of course, t_2 *is* privileged in the sense that at that time it is t_2 and at no other time. However, the same is no less true of t_1. At t_1, t_1 is privileged in the sense that it is t_1 and no other time. And the same is true of every moment. So ultimately all moments are equal.

(30) There is no real problem in translating "X is growing older" into a B-statement. A good translation is provided by 'X is n years old at the time when this token is uttered and is more than n years old at any time later than this time. It is entirely clear that we cannot truly assert this about past things such as Diogenes' cup.

2. See H. N. Castañeda's "Omniscience and Indexical References," *Journal of Philosophy*, 1967, for an interesting discussion of Kretzmann's claim.

(31) The proposed account would be an attempt to exploit the objective difference between time and space, in which the former has direction and the later has no direction. It appears, however, that this difference cannot be responsible for the differences in attitudes toward p_0 and p_2, on the one hand, and toward t_0 and t_2 on the other. After all, it is easy to imagine a world in which space has direction, e. g., in one direction the density of hydrogen atoms increases steadily, or the average temperature increases continually or entropy keeps increasing. Even in such a world it seems clear that if all that we are given is that p_0 is in one direction and p_2 is in the other, we still would not know whether p_0 or p_2 was contemplated with greater apprehension. Thus, it seems that the crucial difference between t_0 and t_2 is not that t_0 is before the t_2 is after t_1, but that t_0 is in the past, i. e., receding from the NOW, and t_2 is in the future approaching the NOW.

(32) There are two problems with this defense. First, whether a is succeeded by b, which is succeeded by c, or a whether succeeds b, which succeeds c, depends on our perspective. Objectively, all we can say is that from the point of view in which a is succeeded by b, b is succeeded by c, but from the point of view in which a succeeds b, b succeeds c. Thus, it is not clear why at t_9 part i should not feel relief that it is a step away from t_{10}, while h is even further at t_8 than g at t_7—yet further—and, of course, by t_1, t_{10} has 'receded' far back into the future.

Secondly, even if Russell were able to explain why, although nothing actually moves along the time series, we have the strong impression of movement, this would not explain why an enlightened person should continue to fear the future once it has been clearly demonstrated to him that in reality nothing fearful is moving toward him. If, for example, after having sighted a mushroom cloud and having heard a tremendous explosion, I form an acute impression that a nuclear explosion has just taken place and dread the effects of the resulting radiation, I then come to learn that no actual explosion has occurred and what I had experienced was merely the visual and audible effects created by a company filming the fictitious events of World War III, I shall cease to worry. Yet. Russellians who claim to have convinced themselves that future events are no more moving toward us than past events do not fear the former any less than other people.

(33) I may well resist the imposition of such an interpretation upon my utterance. I may contend that I am fully resigned to the fact that the later any part of me is located in time the more worn out it is, and I do not wish this to be otherwise. In particular, I am satisfied with the state b is in and I do not wish it to be rejuvenated or changed in any other manner. Yet I may strongly yearn to be ten years younger, which cannot be interpreted in any other manner than that I wish that the NOW located at b should return to a.

(34) I should insist that I realize that any token uttered by b occurs at t_1, which does not worry me at all. I do not wish that any of my temporal parts be elsewhere than where it actually is or that any utterance made by any part of me should occur at a different time than the time at which the particular part making the utterances is located.

CHAPTER III

(35) No, because there are instances in which the opposite is the case: The B-statement 'At no time is parking permitted' entails the A-statement 'Parking is not permitted now', but not vice versa.

(36) the obvious restriction that has to be placed on (ii) is that no component is logically equivalent to p, the original statement.

(37) It is perfectly all right to say that tensed statements are either pure A-statements or what we have been referring to as compounds of A and B-statements. We have defined A-statements as contingent statements which may have different truth-values when asserted on different occasions. Reference to events is not part of the definition. 'John will leave home after Mary returns with the car' is unmistakably a compound of A and B statements, as it may be broken up into three components, two of which are A-statements and one a B-statement.

(38) This is not precise. E does not have and lack the same property. It has the property of being remembered by me at t_1 and what it lacks is a different property, namely, being remembered by me at t_2.

(39)   It is clear that Russell would insist, because of the various arguments cited in the previous chapter, that there are no transient properties in the sense required for both d and e to be true. Thus, here McTaggart seems to yield to the Russellian view. If so, it may be asked, what is ultimately the difference between him and Russell? The answer, as explained at the end of the chapter, is that whereas Russell cheerfully accepts that there are no A-determinations, McTaggart insists that these are absolutely essential for there to be time at all. The discovery that it cannot coherently be maintained that A-determinations exist, for McTaggart, leads to the collapse of the whole notion of time.

(40)   The condition that we benefit from each step we take along a regress is necessary only when the regress is generated through the need to solve a problem or to explain something. In the case of (2) we want to explain what is the cause of E_1, and as soon as we have succeeded in this the need arises to explain the cause of E_2. We say then that in spite of the fact that a new problem arises every time we succeed in explaining a particular event, the regress is not vicious because we will have provided increasingly more solutions to more problems as we advance along the regress. In case of the regress that it is true that a true proposition is true and so on, there is no problem to begin with, and nothing presses on us to take any steps along the regress.

(41)   By making use of Broad's somewhat fanciful idea that beside the regular time series there exists a second-order time series in which every point in the first-order series has an extended history, McTaggart's difficulty could be resolved. The difficulty arose out of the fact that we assigned incompatible properties to the self-same moment M_1. We regarded this as unacceptable, because M_1 is devoid of temporal extension and thus has no room to accommodate incompatible properties. However, we face no difficulty once we are permitted to entertain the possibility of a higher order time series in which M_1 endures indefinitely. All the moments of our regular time series co-exist together at each moment in super-time, and the position of the NOW in regular time varies from moment to moment in super-time. M_1 and every other moment in regular time can assume different properties at different moments in super-time. In particular, M_1 may have the property of futurity, while M_2 has the property of presentness and M_3 the property of pastness at M_1^2, where M_1^2 is a moment in the second-order time series while later at M_2^2, M_1 acquires the property of presentness, M_2

of pastness and M₃ of distant pastness. Thus, the problem of the extensionlessness of the moments in the continuum of which they form a part is resolved with the introduction of a higher order time continuum in which they have unlimited duration.

The reason why this point is of interest is because through it we see that McTaggart's difficulty would disappear under the same circumstances under which the various objections of Russellians to moving NOW disappears. This confirms our view that McTaggart and Russell are ultimately referring to the same difficulty.

CHAPTER IV

(42)   The solution is that (1′) is not a correct translation of (1). From (1) it only follows that:

(1*)   If I am going to be hurt in this air raid then all precautions *I take* are useless.
Similarly, (2′) does not follow from (2) only.

(2*)   If I am not going to be hurt in this air raid then all precautions *I neglect* are unnecessary.
But from (1*) (2*) and (3) it follows only that

(4*)   Either all precautions I take are useless or all precautions I neglect are unnecessary.
But (4*) is not a fatalistic conclusion and is absolutely true.

(43)   There is a crucial difference between the two arguments. Suppose in the first case I do not want P to be true. That will be entirely irrelevant to the question whether P is true or not. In other words let

Wp = I want P to be true.

then ∼ Wp and P together may be true. Consequently, it is quite correct to assert that if P is true, it is not within my power to do S′, and we need not be concerned: but perhaps the truth of P itself in the first place is determined by me? No, the fact that P is true is not brought about by me, for P may be true and Wp false or P false and Wp true.

On the other hand let

W_Q = I want Q to be true.

then $\sim W_Q$ and Q together are never true, for if I do not want there to be a naval battle tomorrow, then I will not issue an order that there be a naval battle but shall order that there be no naval battle, and consequently it is assured that there will be no naval battle tomorrow.

Hence, it is different with (1') than with (1). It does not follow that if Q is true it is not within my power to do O', for it may be claimed that the very question whether to begin with Q is true or not depends on my wishes, for it is not possible for Q to be true and W_Q to be false.

(44) Let us consider the straightforward case in which E_1 at t_1 is said to cause at a later time t_2 a physical event E_2 to occur. Suppose we have a predictor P who already at t_0 is able with complete assurance to predict whether E_2 is or is not going to occur at t_2. It might be asked: Has P predicted correctly at t_0 that E_2 is going to happen at t_2 or that it is not going to happen at t_0? If P predicted correctly that E_2 is not about to happen, then we cannot say that E_2 occurs because of E_1; if P predicted that E_2 was going to occur at t_2, then, as it is given already at t_0 the E_2 at t_2 is the case, the causal effect of E_1 at t_1 is not needed!

In this instance we see at once that the question is misplaced. Where E_1 at t_1 is said to cause E_2 at t_2 obviously E_2 does occur at t_2. Yet we cannot say that because it is given that E_2 occurs at t_2 we do not actually need E_1 at t_1. We need E_1 at t_1, for otherwise it simply would not be the case that E_2 at t_2.

In similar fashion when it is said that X's becoming magnetized at t_1 is caused by it rotating at t_2, X was magnetized at t_1. But of course X at t_2 is needed, for if it was not a fact that X rotates at t_2 it would not be a fact that X would be magnetized at t_1.

(45) No, and what is said on page 121 is not meant to imply this. But if A and B cause C, in the sense that the fact that both A and B obtain is sufficient to ensure that C obtains, then the *statement* that A obtains *implies* that the statement that B obtains implies that C obtains.

(46) We have to distinguish between sufficient and necessary conditions and sufficient and necessary cause. That *a* should be a necessary condition for Tom's decision does not present a problem if we can explain that the ultimate reason for this lies in Tom's decision being a sufficient cause for *a*. Followers of Interpretation F, however, cannot say this and are forced to say that a is a necessary cause and not merely a necessary condition for Tom's decision, and this is what creates a problem.

(47)   Although on page 118 it is said 'I shall not proceed by offering a definition of a cause after reading the series of experiments which lead to the conclusion that α & β is the cause of a, it should become quite obvious that asymmetry in the causal relationship determines which is the cause and which is the effect. When two events are causally connected, an interference with the cause will automatically interfere with the effect as well, whereas we may interfere with the effect as much as we wish without disturbing the cause at all. This characterization of the feature that differentiates a cause from its effect is not new. What is new is the description of the process whereby this feature may become unmistakably discerned. We need to make use of this process, because the feature in question is rather elusive. For example E_1 and E_2 may be causally connected and we may find that we interfere with E_1; this will also have an impact on E_2, yet we cannot infer from this with certainty that E_1 must be the cause. It may be that E_2 is actually the cause, however, for when we thought that only E_1 had been interfered with, in fact, E_2 had also been interfered with: the cause bringing about the interference with E_1—also unnoticed by us—caused an interference with E_2. It is in the course of the kind of experiments we have described that it will gradually become clear which event has been acted on directly and which has been acted on indirectly via an interference with its cause.

(48)   Of course, backward causation cannot be ruled out on the grounds that the past cannot be changed to be different from what, in fact, it has been. After all, when you cause something to happen in the future you do not alter the future to be different from what it is going to be. Similarly, backward causation involves merely changing the past to be different from what it would have been otherwise and causing that it should have been what it was because of what is being done now. Therefore, there has been no reason given as to why my choice now should not determine whether the number on the paper had been prime or composite all along.

(49)   The answer to the first question is that by deriving the contradiction we obtain a reductio proof that some human choices are logically impossible to predict and hence even an Omniscient Being

cannot know ahead of time what they are going to be. All agree that even an Omniscient Being does not know that which is logically impossible for him to know.

As to the second question: if the link between T_i and C_i is absolutely unbreakable then there is simply no point for CH to inquire what for him is best to do, because he is bound to do C_1 if T_1 even though it is crystal clear that C_2 is the better choice.

(50)   We can put to the Perfect Judge this very question: but is it not that ultimately it is preferable for the player to do C_1 for the reason we just gave? And surely his answer would be, no. It is also certain that he cannot be wrong, as he is a completely well-informed judge. Hence, it must be in the best interest of the player to act according to the opinion of the Perfect Judge.

(51)   According to this argument, the predictor who is so extraordinarily endowed as to be able to foresee correctly the choices of the player is nevertheless so defective as not to be able to remember a moment after he has sealed the opaque box that he has not put money in it. It is much more reasonable to assume that he is not a predictor.

To put it differently: according to this argument we would have to maintain the strange hypothesis that a predictor of human choices exists who has a very defective memory. It is hard to see how in practice a defective memory might be a necessary condition for the ability to predict.

(52)   It is not to be denied that on the basis of the data it may be possible to predict with as high probability as you like what the free choices of a certain person are going to be. But on any given occasion a person may decide not to do A, even when A has been designated as the more probable act he will choose. The total incompetence of any predictor manifests itself in this manner: the probability with which he will predict A is not in the slightest affected by the subject's spontaneous decision; that is, he is just as likely to predict A in case the subject contrary to his disposition suddenly decides not to do A as when he does not do so.

(53)   In case T is a necessary cause of A, the occurrence of A is sufficient for the occurrence of T; therefore, if A occurs, it is not true for any reason that T could be prevented. Surely situations are conceivable in which prior to T's occurrence nothing can be done to prevent it.

CHAPTER V

(54) Among the many things that are wrong with this argument one is this: our judgment that a person whose every statement concerning the past is false must fail to understand the notion of 'past' is made in the context of our belief that there has been a past. In other words, in a world in which there has been a past, a person who gets everything wrong about the past cannot have a correct understanding of the notion 'past'. But from this it does not follow that it is the same also in a universe in which there was no past.

(55) It is not inconceivable that a person at this moment should spring into being already in possession of a large store of knowledge. It is not logically necessary to acquire all knowledge through experience; one can be born with knowledge.

(56) 'Confirmability in principle' and meaningfulness are in the same boat as 'confirmed in practice' and credibility, and if the latter need not be abandoned because of the unclarities surrounding the notion of an observation statement, neither need the former. But of course the latter notions are in no trouble, for whatever difficulties there may be in distinguishing accurately between a statement which is an observation statement and a statement which is not, at any time there is a set of statements which is universally regarded as having been established to be credible, irrespective whether they are or are not observation statements. We confirm hypotheses by providing support for them by these established statements. Thus, a sentence is meaningful if a situation can be described which is supported by statements that have been established as true.

(57) It has been assumed for no good reason that the only way in which to confirm p_1 is to actually get hold of a q which can be demonstrated at one and the same time to be highly improbable and true. But of course p_1 may not only be confirmed but also may be rendered firmly credible, even if we have not the slightest idea of what q may be like, simply by showing that p_1 follows either by the rules of deductive or inductive logic from other established propositions. Thus, it is a well-known fact that in any of the many lotteries conducted throughout the world, each of the large number of tickets participating has only an exceedingly small chance of winning the main prize, yet it is absolutely certain even before the drawing takes place that one of them is going to win it. This provides us with many practical situations in which, with respect to any ticket t, the best evidence supports the contention that 't is going to win' is highly improbable, and yet one such proposition is inevitably true. We may treat this as inductive evidence that p_i is true.

In addition, of course, we may also argue from the very nature of the notion of 'probability' that p_i is most likely to be true. After all, p_i is true if any one of $Q_1, Q_2 \ldots Q_n$ is true, where n is an exceedingly large number in view of the fact that there are so many statements about prehistoric times which are most likely to be false. Suppose the probability that Q_1 is true is m and so on. Then the probability that p_i is true is the probability that $Q_1 \vee Q_2 \vee \ldots Q_n$ is true. This, in view of the enormous size of n, is very high even though m is very small.

(58) Far from refuting verificationism, the cited utterance can serve as a useful means to demonstrate the reasonableness of the verificationist position. For it seems clear that if what is intended to be said, that it merely so happens that there is no reason to believe that it is raining, but it would be possible to describe circumstances in which evidence for the rainfall would be available, then indeed the sentence is meaningful, although, of course, it is not in principle unconfirmable. Suppose, however, that what is meant is that it is logically impossible to describe circumstances in which evidence existed that it was or that it was not raining. Then it seems to me that one need not be a logical positivist in order to begin wondering, what sort of a rain are we talking about? It cannot be anything like ordinary rain, which can in principle be seen, heard and felt and whose absence is very strongly confirmed when nothing is seen, heard or felt! It seems evident that in the utterance reference is made to something which we are incapable of describing coherently; that the familiar term 'rain' is not employed at all in a recognizable sense, and thus the sentence has no meaning.

(59) No, of course it may be the case that certain unclarities surround the notion 'coherently describable'. Nobody would suggest, however, that the notion is so unclear as to be utterly vacuous. Nonetheless, it has been concluded by many that the notion of

'verifiable in principle' is so vague as to be entirely useless. Surely then, something very important has been achieved by showing that in fact it is definable in terms of 'coherently describable', a definition which is perhaps not entirely unproblematic yet is of undisputed significance and of frequent use.

Index